Parity and Patterns
in Low Dimensional Topology

Parity and Patterns in Low Dimensional Topology

D.P.Ilyutko, V.O.Manturov and I.M.Nikonov

Cambridge Scientific Publishers

©2015 Cambridge Scientific Publishers

Reviews in Mathematics and Mathematical Physics
Volume 16, Part 1

Printed in UK

ISBN 978-1-908106-47-6 Paperback

Cambridge Scientific Publishers Ltd
45 Margett Street
Cottenham, Cambridge CB24 8QY
UK
janie.wardle@cambridgescientificpublishers.com
www.cambridgescientificpublishers.com

Introduction to the series

In recent years, many young excellent mathematicians both in Russia and abroad made an excellent name for themselves in the international mathematical community. *Reviews in Mathematics and Mathematical Physics* will publish the most outstanding and recent results. Not only new research, but also a detailed background of a problem will be presented to give an insight into a particular direction. This would enable a better grasp of the state of the art in a particular field and an easier positioning of the results discussed.

To maintain high scientific standards, all works published in *Reviews in Mathematics and Mathematical Physics* are peer reviewed by two or three internationally recognized mathematicians. Many of the works presented by Russian mathematicians have been discussed at the seminars at the Faculty of Mechanics and Mathematics, Moscow State University and Steklov Mathematical Institute, Russian Academy of Sciences.

Reviews in Mathematics and Mathematical Physics also plan publications and republications of review papers by outstanding mathematicians to ensure a broader view of the development of modern mathematics and mathematical physics.

A. T. Fomenko

Editor

To our teacher A. T. Fomenko on the occasion of his 70th birthday with
our respect and gratitude

Preface

Various objects in mathematics are equivalence classes of *diagrams* modulo *moves*. For instance, knots are equivalence classes of knot diagrams modulo Reidemeister moves, (finitely presented) groups are equivalence classes of presentations modulo Tietze transformations.

A naïve approach to recognize the equivalence is an attempt to simplify the object by using *decreasing* or at least *non-increasing* moves (with respect to some complexity some complexity). Those moves where one can get to minimal diagrams by a monotonous descent are called *reductive*.

However, in classical knot theory this approach usually fails: in order to unknot a knot, one has to perform various increasing moves.

In the present monograph, we argue that if a diagram is complicated enough then it reveals in all diagrams equivalent to it. This principle works perfectly in virtual knot theory, and the notion of "complicated enough" has to be be justified.

The crucial initial step in this direction was the introduction of the notion of parity: all crossings of a virtual knot diagram are defined to be *even* or *odd* depending on the linkedness of the chord in the respecting chord diagram. Note that for classical knots, all chords of the chord diagram are *even* (the fact already known to Gauß).

However, it turns out that for virtual knots, there exists a natural class of *irreducibly odd* diagrams: oddness guarantees that no decreasing first Reidemeister move can be applied to the diagram, and *irreducibility* is the condition which blocks the possibility to apply any second decreasing Reidemeister moves. Thus, irreducibly odd diagrams are "locally minimal" in the sense that every Reidemeister move applied to them should be increasing. Surprisingly, they turn out to be globally minimal in a very strong sense.

When forgetting over/undercrossing information at classical crossings of virtual knots, one gets to *free knots*; free knots were conjectured trivial by Turaev; it turns out that the parity arguments lead to minimality in a strong sense even for free knots (consequently, for virtual knots), see [47] (cf. [18, 19]).

More precisely, there is a "bracket" $[[K]]$ which takes a free knot diagram K to a $\mathbb{Z}_2$–linear combination of diagrams modulo second Reidemeister moves. Note that the second Reidemeister moves considered separately from other moves are *reductive*, so the equivalence classes modulo second Reidemeister moves are easily recognized just by looking at the minimal representative.

For irreducibly odd diagrams one has $[[K]] = K$, where K, in the right hand side, can be considered just as a graph. This means that the bracket *reduces questions about knots to questions about their diagrams*. Certainly, this does not work for all knots, but this demonstrates another principle: *to simplify the initial objects in order to reduce it to objects with a reductive set of moves*.

In this way, various modifications of the parity principle were found: for odd diagrams where the bracket does not work, one can apply some other formalisms and get similar results which lead to an almost complete simplification.

It turns out that the parity has many applications in classical and virtual knot theory, see [1, 27–29, 35, 37, 47–56]. Using parity a projection from the set of virtual knots to the set of classical knots, invariants of cobordisms of knots were constructed. Despite the fact that parity is trivial in classical knot theory, applications of parity to classical knot theory were found. Also we introduce parity for 2-knots.

Acknowledgments

The authors express their gratitude to L. H. Kauffman, V. A. Vassiliev, A. T. Fomenko, O. Ya. Viro, V. V. Chernov and M. Chrisman for fruitful advices and permanent attention to the work.

The first and third authors were partially supported by grants of RF President NSh – 1410.2012.1, RFBR 12-01-31432, 13-01-00664-a, 13-01-00830-a, 14-01-91161 and 14-01-31288. The second author was partially supported by grants of the Russian Government 14.Z50.31.0020, RF President NSh – 1410.2012.1, RFBR 12-01-31432, 13-01-00830-a 14-01-91161 and 14-01-31288.

Contents

Chapter 1

Basic notions

In this chapter we give basic definitions and notions.

1.1 Classical knots

A (*classical*) *knot* is the image of a smooth embedding of the circle S^1 in the three-dimensional sphere S^3 (or in the space $\mathbb{R}^3$); two knots are called *isotopic*, if one of them can be transformed to the other one by a diffeomorphism of the ambient space S^3 (or $\mathbb{R}^3$) onto itself, preserving the orientation of the sphere S^3 (or $\mathbb{R}^3$) (such a diffeomorphism is called an *isotopy*) (it is well known that each of such diffeomorphisms is homotopic to the identical one in the class of diffeomorphisms [8]). If we embed a disjoint union of several circles $S^1 \sqcup \cdots \sqcup S^1$ in the sphere S^3 (or $\mathbb{R}^3$), then we get a *classical link*; the image of each circle, a knot, is called a *component of the link*. A smooth embedding of a circle (a disjoint union of circles) is also called a *knot* (a *link*). When the orientation of the circle S^1 is fixed, we have an *oriented knot* (respectively, to get an *oriented link* we require orientations for all components). Whenever applying isotopy to an oriented knot (link), this isotopy should agree with the orientation.

A link is *coloured* if its components are enumerated and the isotopy is assumed to preserve the enumeration.

Usually, knots are encoded by plane diagrams, which are 4-valent graphs with an additional structure. Let us define this object.

Let G be a graph with the set of vertices $\mathcal{V}(G)$ and the set of edges $\mathcal{E}(G)$ (we consider only finite graphs, i.e., graphs with finite sets of vertices and edges, and loops and multiple edges are allowed). By an *edge* we mean two half-edges constituting it. We say that a vertex $v \in \mathcal{V}(G)$ has the *degree k* if v is incident to k half-edges. A graph whose vertices have the same degree k is called *k-valent* or a *k-graph*. By abusing notation, it is convenient for us to admit that a *free loop*, i.e., the graph without vertices but with one *cyclic edge*, is a k-valent graph for any k. More precisely, by a *4-graph* we mean the following generalisation of a four-valent graph: a 1-dimensional complex, with each connected component being homeomorphic either to the circle (with no matter how many 0-cells) or to a four-valent graph; by a vertex we shall mean only vertices of those components which are homeomorphic to four-valent graphs, and by edges we mean either edges of four-valent-graph-components or circular components; the latter will be called *cyclic edges*.

Definition 1.1. A 4-graph is called a graph with a *cross structure* or *framed* if for every vertex the four emanating half-edges are split into two pairs of half-edges (we have the structure of opposite edges).

By an *isomorphism* between framed 4-graphs we assume a framing–preserving isomorphism. All framed 4-graphs are considered up to isomorphism. Denote by G_0 the framed 4-graph isomorphic to the circle.

When drawing framed graphs on the plane, we always assume the framing to be induced from the plane $\mathbb{R}^2$.

A diagram corresponding to a knot is constructed as follows. Consider a plane $h \subset \mathbb{R}^3$ (say, $h = Oxy$) and the projection of the knot on it. Without loss of generality, we can assume that the projection of the knot is a finite embedded framed 4-graph being the image of a smooth immersion of the circle in the plane. Speaking informally, a *branch* is a part (i.e., the image of an interval in the three-dimensional space, a part of the circle) of a knot. Sometimes (when it is clear from the context) we also use the term *branch* for the image of a branch in the plane. Each vertex of this graph (also called a *crossing of the diagram of the link*) is endowed with the following additional structure. Let a and b be two branches of the knot, whose projections intersect each other in the point v. Since a and b do not intersect in $\mathbb{R}^3$, the two preimages of v have different z-coordinates. So, we can say, which branch (a or b) goes over (forms an *overcrossing*); the other one goes under (forms an *undercrossing*), see Fig. 1.1. Half-edges of overcrossings are depicted by continuous lines, and half-edges of undercrossings are depicted by lines having a break at the crossing. This image of a knot on the plane is called a *plane knot diagram* or a *knot diagram*.

All crossings of a diagram of an oriented link are divided into *positive* ones $\times$ and *negative* ones $\times$. It is easy to check that in the case of a knot the sign of a crossing does not depend on an orientation of a knot.

A link diagram is called *alternating*, if overcrossings and undercrossings alternate when going along each component.

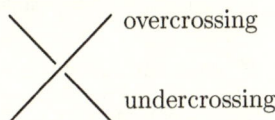

overcrossing

undercrossing

Fig. 1.1   The local structure of a crossing

The simplest examples of diagrams of classical knots are depicted in Fig. 1.2. The first two diagrams represent the *unknot* or the *trivial knot*, the third diagram represents the *trefoil*, and the knot shown in the fourth diagram is called the *figure eight knot*. All these diagrams are alternating.

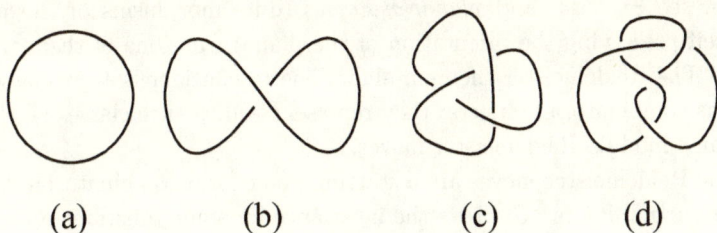

(a)   (b)   (c)   (d)

Fig. 1.2   The simplest knots

In Fig. 1.3(a) the *trivial link with two components* is depicted. In Figs. 1.3(b), 1.3(c) and 1.3(d) the *Hopf link*, the *Whitehead link* and the *Borromean rings* are shown, respectively. Each of these links is not trivial.

By an *arc* of a planar link diagram we mean a connected component

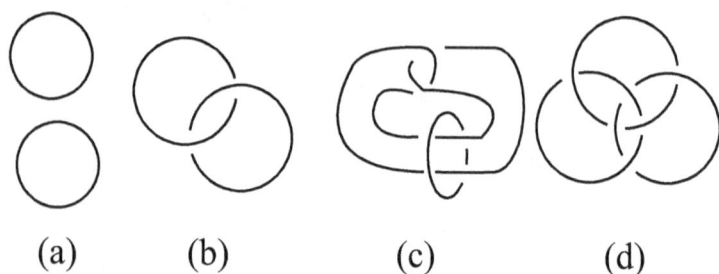

(a)   (b)   (c)   (d)

Fig. 1.3   The simplest links

of the diagram (thus, each arc always goes "over"; it starts and stops at undercrossings).

Definition 1.2. The framed 4-graph without an over/undercrossing structure is called the *shadow* of the knot. The *complexity* of a knot is the minimal number of crossings for knot diagrams of a given isotopy type. We shall call a link diagram *connected* if its shadow is connected as a 4-graph. In particular, any knot diagram is connected.

Reidemeister [61] proved that any two planar diagrams generated isotopic knots (links) if and only if there existed a finite chain of *Reidemeister moves*, see Fig. 1.4, and *planar isotopies* (diffeomorphisms of the plane on itself preserving the orientation of the plane) from one of them to the other. The Reidemeister theorem allows one to consider isotopy classes of links as combinatorial objects: they represent equivalence classes of planar diagrams modulo Reidemeister moves.

The Reidemeister moves are a starting point for a combinatorial definition of virtual knots. To prove the invariance for some functions on knots, one should check its invariance under the Reidemeister moves.

Since Gauß time it has been known that knots can be encoded by chord diagrams with certain additional information. If we view a knot diagram as an immersion of the circle into the plane, we can obtain a *chord diagram* by joining those points on the circle which are mapped to the same point under the immersion, by chords.

Definition 1.3. A *Hamiltonian cycle* on a graph is a cycle passing through

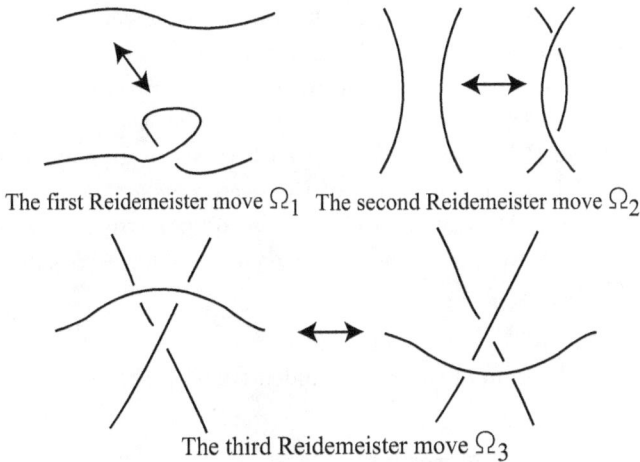

The first Reidemeister move Ω_1 The second Reidemeister move Ω_2

The third Reidemeister move Ω_3

Fig. 1.4   Reidemeister moves $\Omega_1, \Omega_2, \Omega_3$

all vertices of the graph. By a *chord diagram* we mean a cubic graph consisting of one selected non-oriented Hamiltonian cycle (*core circle* or *circle*) and a set of non-oriented edges (*chords*) connecting points on the cycle, moreover, distinct chords have no common points on the cycle.

We say that two chords of a chord diagram are *linked* if the ends of one chord belong to two different connected components of the complement to the ends of the other chord in the core circle. Otherwise, we say that chords are *unlinked*.

Remark 1.1. As a rule, a chord diagram is depicted on the plane as the Euclidean circle with a collection of chords connecting end points of chords (intersection points of chords which appear as artifacts of drawing chords do not count as vertices).

We shall also deal with *oriented chord diagrams*, i.e., chord diagrams with its core circle to be oriented. In that case we consider chord diagrams up to isomorphisms of graphs preserving the orientation of the core circle.

Having a classical knot diagram one can assign to it a chord diagram with an additional structure, the *Gauss diagram* [59].

Definition 1.4. Let K be a planar diagram of an oriented classical knot. Let us fix a point on the diagram distinct from a vertex of the diagram. The *Gauss diagram* $G(K)$ corresponding to K is the chord diagram consisting of the core circle (with the fixed point) on which the preimages of the overcrossing and the undercrossing for each crossing are connected by an arrow oriented from the preimage of the overcrossing to the preimage of the undercrossing. Moreover, each arrow is endowed with a sign equal to the sign of the crossing, i.e., the sign is equal to 1 for a crossing $\times$ and -1 for a crossing $\times$.

The Gauss diagram of the right-handed trefoil is shown in Fig. 1.5.

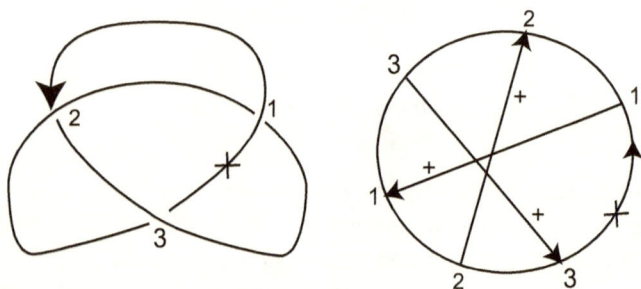

Fig. 1.5   The Gauss diagram of the right-handed trefoil

It is not difficult to rewrite the Reidemeister moves in the language of Gauss diagrams. As a result, we can think of a knot as an equivalence class of Gauss diagrams modulo formal moves.

If we consider an arbitrary chord diagram and try to depict a classical knot corresponding to it (for any orientation and signs of chords), then, in many cases, we shall fail, see [4, 5, 46, 60]. Let us consider the chord diagram D depicted in Fig. 1.6. It is easy to see that there is no knot diagram such that Gauss diagram of which is D for whatever choice of orientations and signs of chords [32].

Admitting arbitrary chord diagrams (with arrows and signs) as Gauss diagrams and the Reidemeister moves as moves on all chord diagrams, we get a new theory, the *theory of virtual knots* [22].

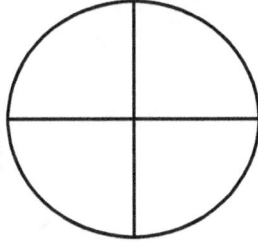

Fig. 1.6 "Non-realisable" chord diagram

1.2 Virtual knots

In [32] Louis Kauffman introduced the notion of a *virtual knot* (or, in the case of many components, a *virtual link*). A virtual knot represents a natural combinatorial generalisation of a classical knot: we introduce a new type of a crossing and add new moves to the list of the Reidemeister moves. The new crossing (called *virtual* and depicted by a small circle) should be treated neither as passage of one branch over the other nor as a passage of one branch under the other. It should be treated as a diagrammatic picture of two parts of a knot (a link) on the plane which are far from each other, and the intersection of these parts is artifact of such a drawing. In that sense the following list of *generalised Reidemeister moves* is natural: All classical Reidemeister moves related to classical crossings and a *detour move*. The latter represents the following: A branch of a knot diagram containing several consecutive virtual crossings but not containing classical crossings, can be transformed into any other branch with the same endpoints; new intersections and self-intersections are marked as virtual crossings, see Fig. 1.7.

Definition 1.5. A *virtual diagram* (or a *diagram of a virtual link*) is the image of an immersion of a framed 4-graph in $\mathbb{R}^2$ with a finite number of intersections of edges. Moreover, each intersection is a transverse double point which we call a *virtual crossing* and mark by a small circle, and each vertex of the graph is endowed with the classical crossing structure (with a choice for underpass and overpass specified). The vertices of the graph with that additional structure are called *classical crossings* or just *crossings*.

A virtual diagram is called *connected* if the corresponding 4-graph is

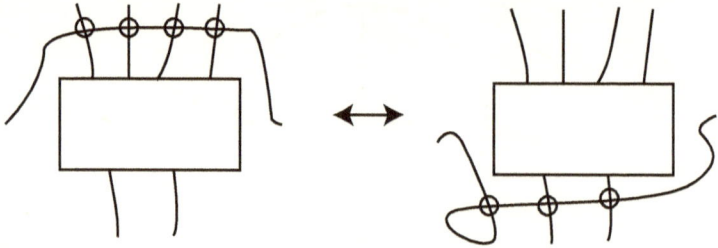

Fig. 1.7   Detour move

connected.

A *virtual link* is an equivalence class of virtual diagrams modulo the generalised Reidemeister moves. The latter consist of the usual Reidemeister moves referring to classical crossings and the detour move.

Thereby, in order to define a virtual knot we should know only the positions of classical crossings and their connections with each other. Moreover, positions of paths connecting classical crossings, their intersections and self-intersections, are not important for us.

One can naturally define *components* of a virtual link. We can define components of a link combinatorially. Namely, by a *unicursal component* of a virtual link diagram we mean the following. Consider a virtual diagram as a one-dimensional cell-complex on the plane. Some of the connected components of this complex are circles: we call each such component a (*unicursal*) *component* of a link. The remaining part of the cell-complex represents a 4-graph with vertices which are classical or virtual crossings. *Unicursal components* of the diagram are (besides circles) equivalence classes on the set of edges of the graph: two edges e, e' are *equivalent* if there exists a collection of edges $e = e_1, \ldots, e_m = e'$ and a collection of vertices $v_1, \ldots, v_{m-1}$ (some of them may coincide) of the graph such that edges e_i, e_{i+1} pass to the vertex v_i from the opposite sides.

A *virtual knot* is a virtual link with one *unicursal component*.

The *writhe number* $w(K)$ of a virtual diagram K is the number equal to the number of positive crossings $\diagup\!\!\!\!\diagdown$ minus the number of negative crossings $\diagdown\!\!\!\!\diagup$.

It is easy to see that the number of components of a virtual diagram is invariant under the generalised Reidemeister moves. In the classical case this definition coincides with the definition given before.

Remark 1.2. Note that such approach, the standard moves inside a local Euclidean domain and the detour move, was used by N. Kamada and S. Kamada [31] for constructing formal theories of multidimensional "virtual knots" and their invariants.

Branches of a virtual link having a virtual intersection and related to two parts of a virtual link diagram located far away from each other, can freely move on a surface independently from each other. This leads us to a definition of virtual links as links in thickened oriented surfaces $S \times I$, where S is a two-dimensional oriented closed surface and $I = [0, 1]$ is a segment with a fixed orientation; moreover, thickened surfaces should be considered up to stabilisations, i.e., up to additions and removals of handles to S in such a way that additional thickened handles do not touch our link.

From now on, we suppose that the structure of direct product is fixed on a thickened surface $S \times I$ and it is pointed toward which side is up, the side $S \times \{1\}$, and which one is down, the side $S \times \{0\}$.

In the case of links, one can consider a disjoint union of manifolds, $S_1 \sqcup \cdots \sqcup S_k$ (sometimes it is required that in each manifold $S_j \times I$ there is at least one component of a virtual link, [38]). Links in $S \times I$ are described by diagrams on S with the over/undercrossing structure specified. In that sense virtual diagrams are obtained by means of regular generic projections of diagrams from S to the plane: crossings pass to classical crossings and new intersections (artifacts) are marked by virtual crossings; moreover, it is required that under regular generic projections neighborhoods of classical crossings are mapped into the oriented plane with the orientation preserved. The Reidemeister moves for diagrams on S (the same as in the case of classical diagrams) correspond to the classical Reidemeister moves for virtual diagrams; there are also transformations which do not change the combinatorial structure of a diagram on S, but do change the combinatorial structure of the projection to the plane: the detour move corresponds to these transformations.

The theorem about an equivalence of different definitions of virtual knots was announced in the work [33] and proved by different authors, including Kauffman.

Let us describe the construction from [7]

Let K be a virtual diagram, and let S be a closed oriented 2-surface. We call the pair $P = (S, K)$ *a canonical link surface diagram* (CLSD) if there exists an embedding of the underlying framed 4-graph of K into S such that the complement to the image of this embedding is a disjoint union of 2-cells. Denote by $\widetilde{S}$ a neighborhood of the embedding of K in S. For a CLSD, $P = (S, K)$, if there exists an orientation preserving embedding $f \colon \widetilde{S} \to M$ into a closed oriented surface M, we call $f(K)$ *a diagram realisation* of K in M. Two CLSD's $P = (S, K)$ and $P' = (S', K')$ are related by *an abstract Reidemeister move* if there is a closed oriented surface M and diagram realisations of K and K' in M which are related by a Reidemeister move in M. Two CLSD's are *equivalent* if they are related by a finite sequence of abstract Reidemeister moves. Following N. Kamada and S. Kamada [31] one can construct a bijection

$$\psi \colon \{\text{virtual link diagrams}\} \to \{\text{CLSD's}\}.$$

The idea of this map is illustrated in Fig. 1.8. Having a virtual link diagram K, we take all classical crossings of it and associate two crossing bands (a piece of 2-surface) with a neighborhood of a crossing, and a pair of skew bands (when drawing on the plane it does not matter which band is over and which one is under) with a virtual crossing. If we connect these crossings and bands by (non-overtwisted) bands going along edges, we get a 2-surface with boundary. Gluing its boundary components by discs, we get an orientable closed 2-surface. We call $\psi(K)$ *a CLSD associated with a virtual diagram K*.

Fig. 1.8   The local structure

A realisation of the detour move by moves on thickened surfaces and their projections is shown in Fig. 1.9.

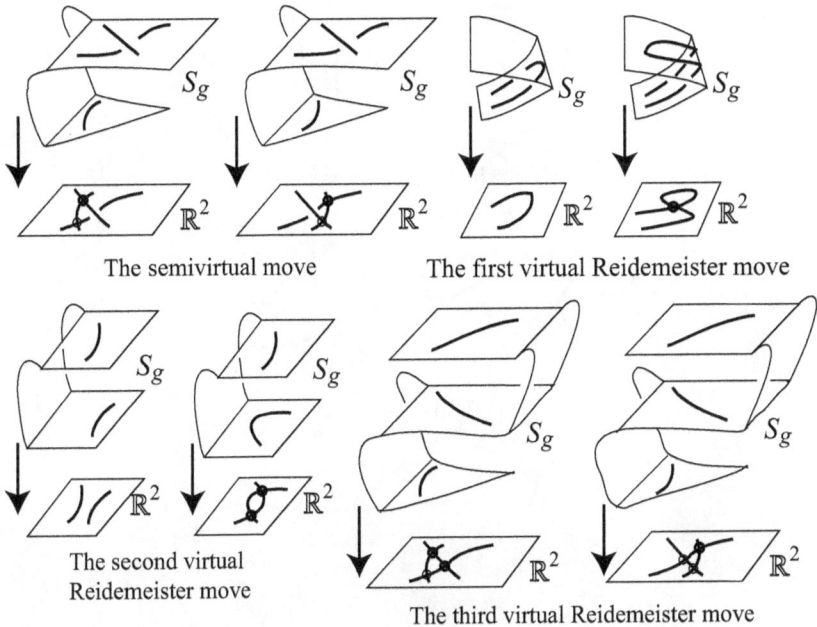

The semivirtual move   The first virtual Reidemeister move

The second virtual
Reidemeister move

The third virtual Reidemeister move

Fig. 1.9   Generalised Reidemeister moves and thickened surfaces

This leads us to local versions of the detour move which consist of:

(1) *Virtual Reidemeister moves* $\Omega'_1, \Omega'_2, \Omega'_3$, which are obtained from the classical Reidemeister moves by swapping all classical crossings participating in moves for virtual crossings, see Fig. 1.10.

(2) *Semivirtual version* Ω''_3 of the third Reidemeister move. Under this move the branch containing two virtual crossings can be carried through a classical crossing, see Fig. 1.11.

We call a Reidemeister move *increasing* (respectively, *decreasing*) if this move increases (respectively, decreases) the number of crossings (the number of classical crossings in the classical case and the number of virtual ones in the virtual case). For example, the moves $\Omega_1, \Omega_2, \Omega'_1, \Omega'_2$ are increasing "in one direction", and decreasing "in the opposite direction".

The following statement is obvious.

The move Ω'_1 The move Ω'_2

The move Ω'_3

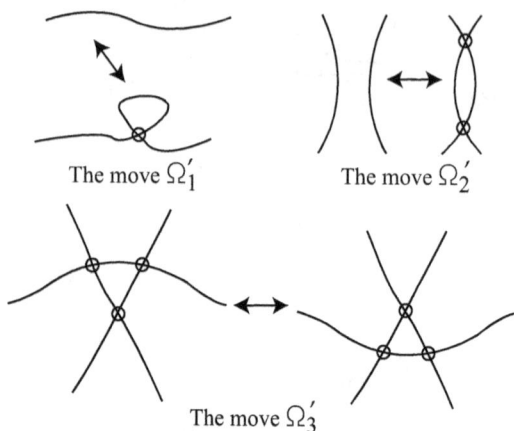

Fig. 1.10   Moves $\Omega'_1, \Omega'_2, \Omega'_3$

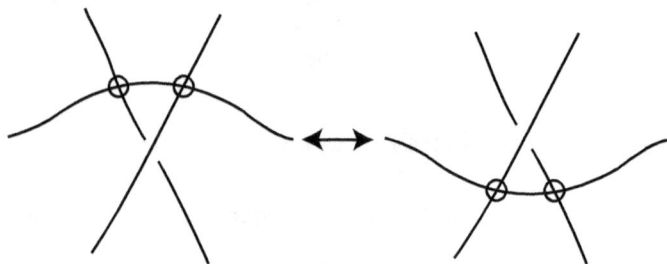

Fig. 1.11   The semivirtual move Ω''_3

Statement 1.1. *Two virtual diagrams K and K' are obtained from each other by a finite sequence of detour moves if and only if they are obtained from each other by a finite sequence of moves Ω'_1, Ω'_2, Ω'_3, Ω''_3 and their inverses.*

The reconstruction of a knot diagram in a thickened surface by a virtual diagram on the plane goes as follows.

Let K be a virtual diagram on the sphere S^2 (we compactify the plane $\mathbb{R}^2$ by adding one point). Each virtual crossing of this diagram corresponds

to an intersection of two arcs. Let us choose one of them and construct a
handle for its "lifting", see Fig. 1.12. As a result, we get a diagram (with
over- and undercrossings and virtual crossings) on the torus, the number of
virtual crossings of which is less by one than the number of virtual crossings
in the initial diagram.

Note that the choice for a position of a handle, up or down, is immaterial
since thickened surfaces are considered on their own account without any
embedding into $\mathbb{R}^3$.

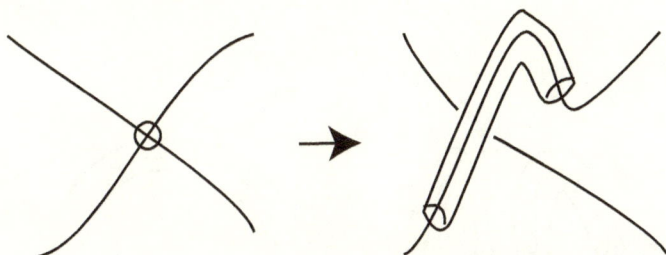

Fig. 1.12   Lifting of a virtual crossing to a handle

It is also easy to check that it does not matter which arc we choose for
lifting it to a new handle. Two diagrams K_1 and K_2 corresponding to two
such lifts to surfaces M_1 and M_2 with handles, i.e., $K_1 \subset M_1$ and $K_2 \subset M_2$, will be combinatorially equivalent (i.e., there exists a homeomorphism
$f \colon M_1 \to M_2$ of one lift to the other one and f transforms one virtual
diagram to the other one, $f(K_1) = K_2$).

Continuing this process, we can get rid of all virtual crossings and obtain
a diagram on S_g ($=S_g \times \{\frac{1}{2}\} \subset S_g \times [0,1]$), where g is the number of
handles. Here it is convenient to use detour moves. Each of these moves is
a "merging" of subsequently situated handles to one handle and partition
this handle into new handles, situated in other places, see Fig. 1.13.

In Fig. 1.13 (lower part) the "merging" (respectively, the partition)
consists of elementary moves which are the destabilisation (respectively,
the stabilisation). Meanwhile, classical Reidemeister moves are performed
locally on some part of the surface S_g obtained from the sphere by adding
handles.

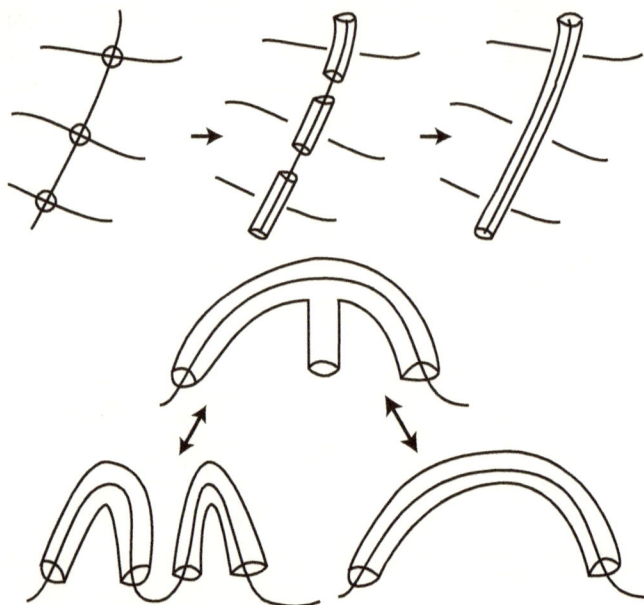

Fig. 1.13   Detour move and stabilisation

It is natural that the surface S_g is automatically oriented. The orientation for S_g arises from the orientation of the sphere S^2 to which we attach handles.

Note that on the surface S_g there is no fixed system of longitudes and meridians. Actually, under first virtual Reidemeister move this surface goes through a *Dehn twist*, see Fig. 1.14.

Note that two moves resembling Reidemeister moves and shown in Fig. 1.15, generally speaking, are not equivalence relations on virtual knots. They are called the *forbidden moves*, see Fig. 1.15.

It turns out (it was noted independently by Goussarov, Polyak, Viro [22] and Kauffman [32, 33]) that if we admit two forbidden moves, then any two virtual knots are equivalent (this assertion is not true for links). If we add only one of these moves and the other one is left forbidden, we obtain an interesting theory of *welded knots*. This theory was proposed by Satoh [62], see also [35].

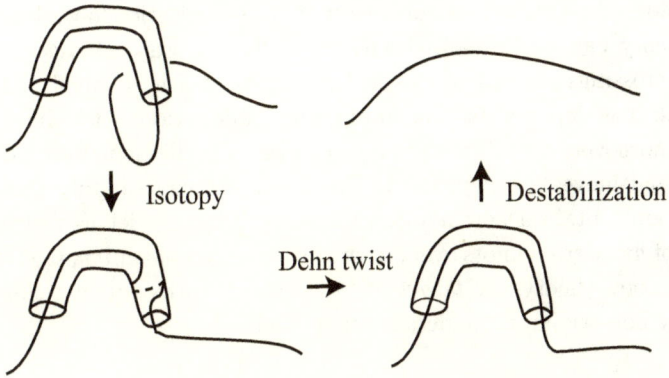

Fig. 1.14 Dehn twist and the move Ω'_1

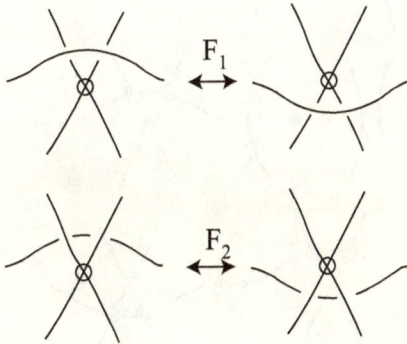

Fig. 1.15 Forbidden moves

It is well known (a proof can be found, e.g., in [44]) that each classical knot can be transformed into the unknot by a subsequent change of the crossing structure $\times \Longleftrightarrow \times$. This is a starting point for constructing knot invariants (skein-modules, Conway algebra, Kauffman polynomial, Vassiliev invariants and so on). This is not true for virtual knots.

Namely, if we factorize the theory of virtual knots over the relation allowing to change the structure of classical crossings, we get the nontrivial theory of *flat virtual knots*, see, e.g., [31] (flat virtual knots can be considered as curves on surfaces modulo stabilisation/destabilisation). This theory can be formalised with the following way. We use only one sort of crossings instead of classical crossings, which is called *flat* or *flat classical*; it is depicted just as an ordinary intersection of two lines on the plane; moreover, we allow virtual crossings. The Reidemeister moves for flat virtual moves are depicted in Fig. 1.16. We can simplify this theory further and obtain a new non-trivial theory. Namely, let us factorize the theory of flat virtual knots over a new move, the *virtualisation*, see Fig. 1.17. We get a new theory, the *theory of free knots*. It turns out that this theory is highly non-trivial, more details see in Chap. 2.

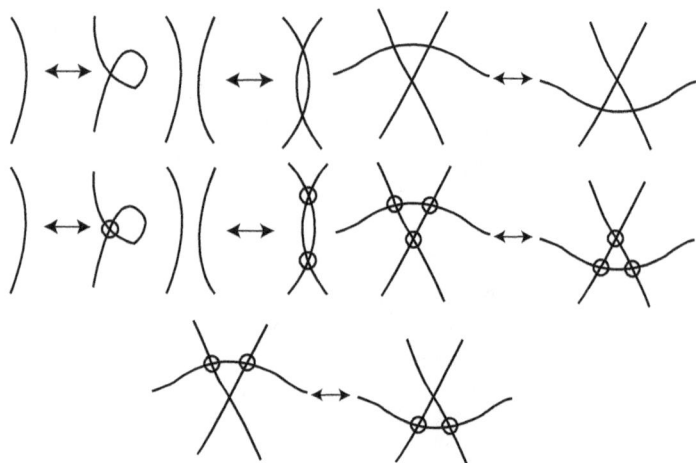

Fig. 1.16   Reidemeister moves for flat knots

Another simplification of the theory of virtual knots when we keep in mind only the local writhe information (at each classical crossing) is called the *theory of pseudoknots*. But if we keep in mind the information about which branch lies upper and which one is lower, we obtain the theory called the *theory of quasiknots*, see [68].

The simplest example of a flat virtual knot not equivalent to the unknot is depicted in Fig. 1.18.

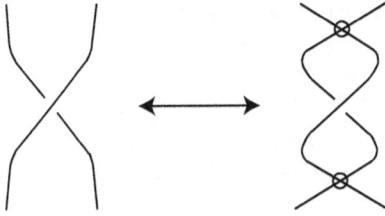

Fig. 1.17 Virtualisation

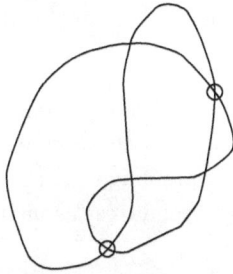

Fig. 1.18 The simplest non-trivial flat virtual knot

We know that virtual knots can be lifted to thickened surfaces. If we disregard which branch of a virtual knot in a classical crossing passes over and which one passes under (i.e., we forget the over/undercrossing structure at classical crossings), we shall obtain a natural lifting of flat virtual knots to two-surfaces ("thickening" is not required). This leads us to the theorem, see, e.g., [31].

Theorem 1.1. *Flat virtual links are equivalence classes of finite collections of curves in 2–surfaces up to free homotopy, stabilisation and destabilisation.*

The problem of whether two such representations of a flat virtual link are equivalent in the category of flat virtual knots is algorithmically recognised.

Let S be an oriented closed 2-surface of genus g. In [24] it is showed that if we had two homotopic curves on the surface S such that the numbers of self-intersection points for these curves were minimal, then these two curves could be related to each other by only a finite sequence of third Reidemeister moves.

Theorem 1.2 (see [28]). *A flat knot has a unique minimal (with respect to genus) representative (S, K) (closed oriented 2-surface, a curve on it) up to a diffeomorphism $(f : (S, K) \to (S', K'),\ K' = f(K))$ and a homotopy of knots on the surface.*

As a corollary from this theorem we get the following statement.

Corollary 1.1. *A flat knot has a unique minimal representative (with respect to genus and the number of crossings) up to a diffeomorphism and third Reidemeister move.*

Not all Gauss diagrams are realisable by embeddings of framed 4-graphs in the plane, but they can be realised as *generic immersions* by marking points having more than one preimage (in the case of generic immersion these points have exactly two preimages) as virtual crossings, see Fig. 1.19.

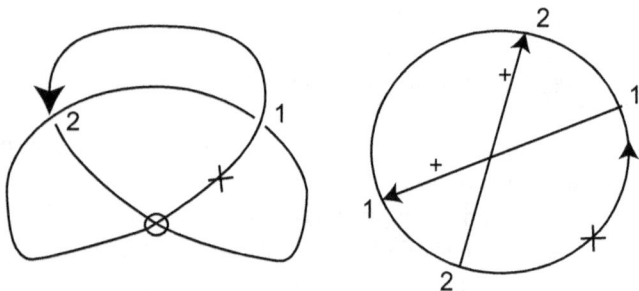

Fig. 1.19   Gauss diagram of a virtual knot

This naturally leads us to the following definition of a virtual knot (not a link): one has to consider all formal Gauss diagrams and describe formal Reidemeister moves on them (as it was done in the case of classical knots): they will represent combinatorial schemes of transformations of Gauss diagrams. In this case equivalence classes of Gauss diagrams modulo formal Reidemeister moves are virtual knots. Note that we do not need the detour move, since a Gauss diagram "knows" nothing about positions of virtual crossings on the plane, and it "knows" only positions of classical crossings and their connections with each other. It means that Gauss diagrams "feel" only classical Reidemeister moves and do not "feel" the detour move.

1.3 Atoms and knots

The notion of *atom* was first introduced by Fomenko in [17] (see also [3, 13–16]) for the classification purposes of integrable Hamiltonian systems of low complexity. A connection between atoms and knots was investigated by the second-named author in the series of works [41–44].

In [64] Turaev was using a similar construction (hence, the atom genus is also called the *Turaev genus* [39]).

Definition 1.6. An *atom* is a pair (M, Γ): a connected closed 2-manifold M and an embedded 4-valent graph $\Gamma \subset M$ such that $M \backslash \Gamma$ is a disjoint union of cells that admits a checkerboard colouring (with black and white colours), i.e., two components of $M \backslash \Gamma$ being adjacent by an edge of Γ have always distinct colours.

The graph Γ is said to be the *frame* of the atom. A *vertex* of the atom is a vertex of this frame.

The *genus* (respectively, *Euler characteristic*) of the atom is that of its first component, i.e., the surface. An atom is called *orientable* (respectively, *connected*) if the corresponding manifold is orientable (respectively, connected). An atom of genus zero is called *spherical*, oriented atoms of genus one are called *torus*.

The example of a torus atom is depicted in Fig. 1.20.

If an atom is orientable, then we can equip it with an orientation. Thus, it makes sense to say about oriented and unoriented atoms.

Atoms are considered up to the natural isomorphism. Two atoms are called *isomorphic* if there exists a one-to-one map of their first components taking frame to frame and black cells to black cells.

Each atom (more precisely, its equivalence class) can be completely restored from the following combinatorial structure:

1) the frame (a 4-valent graph);
2) the *A-structure* or the *structure of opposite edges*;

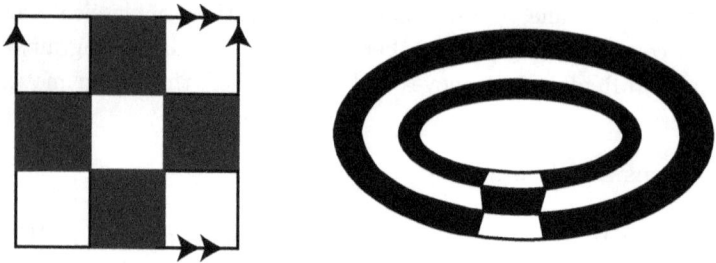

Fig. 1.20   A torus atom

3) the *B-structure* (for each vertex, we indicate some two pairs of adjacent half-edges (also: two angles) which constitute a part of the boundary of black cells).

Atoms are very convenient for describing links and virtual links.

Given a virtual diagram K (each connected component of the underlying graph of K has at least one classical crossing), let us construct the atom $\mathrm{At}(K)$ associated with K. Vertices of $\mathrm{At}(K)$ are in one-to-one correspondence with classical crossings of the diagram K. Classical crossings of K are connected to each other by branches of K which may intersect each other only at virtual crossings. At each classical crossing we have four emanating half-edges e_1, e_2, e_3, e_4 in clockwise-ordering such that the pair (e_1, e_3) forms an undercrossing and the pair (e_2, e_4) forms an overcrossing. These edges are in one-to-one correspondence with edges of $\mathrm{At}(K)$ connecting the corresponding vertices. The 1-cycles of the frame pasting black and white cells are constructed as follows. Each boundary of a 2-cell is a rotating circuit on the frame, i.e., a circuit which passes every edge at most once and switches at each vertex from an edge to an adjacent (non-opposite) one. Black cells are glued to the angles formed by (e_1, e_2) and (e_3, e_4), and white cells are glued to the angles formed by (e_2, e_3) and (e_1, e_4).

Define the *genus*, cf. [63], of a virtual link as the minimum of values $g(M, \Gamma)$ over all atoms (M, Γ) corresponding to diagrams of the link, and the *height* as the minimum of values $h(M, \Gamma)$ over all atoms (M, Γ) corresponding to diagrams of the link.

From the definition it follows that classical links have height zero. Actually, if a link is classical, then its shadow is a framed 4-graph embedded in the plane which is the frame of the corresponding atom. On construction the A-structure of the atom coincides with the A-structure induced from the plane. Thus, *a height atom is assign to each classical diagram.*

Note that the inverse procedure of constructing a virtual diagram from an atom is not well defined, for we get a virtual diagram up to detour moves and virtualisations, see, e.g. [44]. The latter follows from the fact that the framing of the frame of an atom does not give a cyclic structure on outgoing half-edges: Knowing that the half-edges e_1, e_3 are opposite to each other, we have two possible cyclic orders for embedding a neighborhood of this vertex into the plane: e_1, e_2, e_3, e_4 and e_1, e_4, e_3, e_2. The diagrams corresponding to them differ from each other by (detour moves) and virtualisations.

Having an atom (M, Γ), we can construct an oriented diagram of a virtual link as follows. Let us consider a generic immersion of the frame Γ in the plane with the A-structure preserved. The image of this immersion will be a framed 4-graph, vertices of which are images of atom vertices and intersections of interior points of edges. The latter will be considered as virtual crossings, and images of vertices as classical crossings. For restoring the structure of classical crossings we shall use the B-structure of the atom. Namely, an edge forms an overcrossing if while moving inside an angle of the supercritical level in clockwise ordering, it moves from an undercrossing to an overcrossing edge.

Denote all possible classes of virtual links obtained from an atom At by $L(\mathrm{At})$.

An ambiguity appears when an immersion is specified, see Theorem 1.4.

In the case of height atoms one can restrict by considering embeddings. Denote the corresponding subset of the set $L(\mathrm{At})$ by $L_{\mathrm{emb}}(\mathrm{At})$.

We have the following theorem.

Theorem 1.3 (see [43]). *The isotopic type of a link from $L_{\mathrm{emb}}(\mathrm{At})$ does not depend on the embedding of the frame of the atom At into the plane with the A-structure preserved.*

Note that with such an approach the tautological embedding of flat atoms into the plane is associated with alternating diagrams; they divide the plane (the sphere) into black and white cells.

In the case of immersions there is no well definedness.

Theorem 1.4 (see [44, 56]). *Diagrams of links from the set $L(\mathrm{At})$ constructed from an atom At are obtained from each other by applying the detour move and virtualisation.*

Let an atom be given. Assume that for the A-structure of the atom there exists an orientation of all edges of the atom such that at each vertex two opposite edges are emanating and two other opposite edges are coming.

Definition 1.7. We call this structure the *source–sink* structure.

Proposition 1.1. *The frame of an atom admits a source–sink structure if and only if the atom is orientable.*

The following theorem holds.

Theorem 1.5 (see [56]). *The atom $\mathrm{At}(K)$ is orientable if and only if all chords of the diagram $G(K)$ are even.*

In particular, the property of an atom to be orientable depends only on the free knot corresponding to K, and does not depend on over/undercrossing information at classical crossings of K.

Example 1.1. Let us consider the Gauss diagram depicted in Fig. 1.21 (the left part). An orientation of arcs is given in it, this orientation gives rise to a source–sink structure for the frame of the corresponding atom. It is not difficult to check that one of the atoms corresponding to this diagram is spherical (i.e., the surface of the atom is the sphere), therefore, it is orientable, i.e., all atoms with the same frame are orientable.

We cannot define a source–sink structure for the chord diagram drawn in Fig. 1.21 (the right part), since the chord a is odd (it is linked with one chord). Therefore, an alternating orientation of arrows along the core circle of the chord diagram leads to four incoming edges for the chord a.

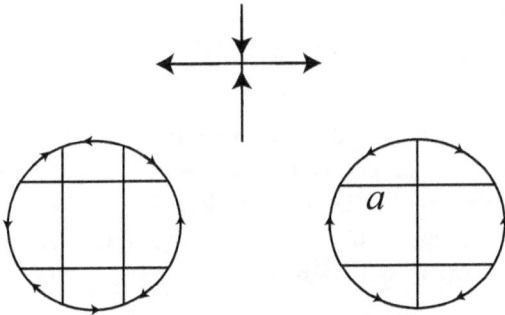

Fig. 1.21   Gauss diagrams and the source–sink structures

1.4   Kauffman bracket and self-linking number

Let us give two examples of invariants of virtual knots. The first of them is
the Kauffman bracket polynomial. In the following chapters we refine this
invariant by using a notion of parity, and generalise it for new combinatorial
objects.

The second invariant, described below, belongs to Kauffman and is
based on the notion of parity. This invariant is the first one in which
parity is used. As we shall see later (Chap. 2), the genuine notion of parity
gives deep, important and interesting results and insights.

Let K be a non-oriented virtual link diagram with n classical crossings.
Each crossing of K can be *smoothed* in one of the two ways: the positive
smoothing A: $\times \to)($ or the negative smoothing B: $\times \to \asymp$. A
choice of a smoothing at each classical crossing is called a *state* of the
diagram. Thus, the diagram K has 2^n possible states. After a smoothing
of a diagram with respect to some state we obtain a virtual link diagram
without classical crossings. Thus, the obtained virtual link is trivial. With
each state s we associate the following three numbers: $\alpha(s)$ is the number
of classical crossings in the state s, having A–smoothing , $\beta(s) = (n - \alpha(s))$
(the number of B–smoothings in the state s), and $\gamma(s)$ is the number of
circles of the (trivial) link in the state s.

The *Kauffman bracket polynomial* [33] $\langle \cdot \rangle$ of the non-oriented diagram

K is defined by the formula:

$$\langle K \rangle = \sum_s a^{\alpha(s)-\beta(s)}(-a^2 - a^{-2})^{(\gamma(s)-1)},$$

where the sum is taken over all states s of the diagram K.

The Kauffman bracket polynomial is invariant under all generalised Reidemeister moves except for the first classical Reidemeister move.

From the definition it follows that the Kauffman bracket polynomial satisfies the following axiomatic relation

$$\langle \,\times\, \rangle = a \langle\,)(\, \rangle + a^{-1} \langle\, \asymp\, \rangle.$$

Definition 1.8. The difference between the leading term degree and the degree of the lowest non-zero term of a Laurent polynomial $P(x)$ is called the *span* of $P(x)$ and is denoted by span $P(x)$.

Note that the span of the Kauffman bracket polynomial is invariant under all Reidemeister moves.

The Kauffman bracket polynomial which is "almost" invariant for non-oriented virtual links leads us to a construction of an invariant of oriented virtual links. For constructing this invariant we should normalise the Kauffman bracket polynomial. Let K be an oriented virtual diagram. Set

$$X(K) = (-a)^{-3w(K)} \langle |K| \rangle,$$

where $w(K)$ is the writhe number of K, and $|K|$ is the non-oriented diagram obtained from K by forgetting the orientation.

It is known [33] that we get an invariant polynomial (a Laurent polynomial) of oriented links, which is called the *Jones–Kauffman polynomial* [30]. Among the properties to be partially detected by the Jones–Kauffman polynomial we emphasise the following:

(1) Is a link classical?
(2) What is the genus of a link (in the sense of atoms)?
(3) What is the number of crossings of a given link diagram (lower estimate)?

Note that some properties which can be difficult detected by the Jones polynomial are easily detected by using parity.

Let us construct the second invariant.

Definition 1.9. Let K be a diagram of an oriented virtual knot. Call a classical crossing of K *odd*, if in the Gauss diagram of K the number of chords linked with the corresponding chord is odd.

Set:

$$J(K) = w(K)|_{\mathrm{Odd}(K)},$$

where $\mathrm{Odd}(K)$ denotes the collection of odd crossings of K, and the restriction of the writhe number $w(K)$ to $\mathrm{Odd}(K)$, denoted by $w(K)|_{\mathrm{Odd}(K)}$, means the summation of the signs of the odd crossings in K. Then it is not hard to see that $J(K)$ is an invariant of the virtual knot (link) K. We call $J(K)$ the *self-linking number* of the virtual diagram K. This invariant is simple, but surprisingly powerful.

If K is classical, then $J(K) = 0$, since all crossings in any classical diagram are even.

Theorem 1.6. *Let K be a virtual knot diagram, and let K^* denote the mirror image of K (obtained by switching all the crossings of the diagram K). Then*

$$J(K^*) = -J(K).$$

Hence, if $J(K) \neq 0$, then K is inequivalent to its mirror image. If K is a virtual knot and $J(K)$ is non-zero, then K is not equivalent to any classical knot.

We leave the proof of this theorem and the proof of the invariance of $J(K)$ to the reader. See [34] for more about this invariant.

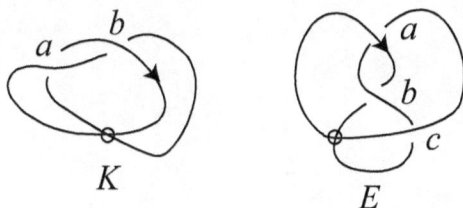

Fig. 1.22   Virtual trefoil K and virtual figure eight E

In Fig. 1.22, the two virtual knots illustrate an application of Theorem 1.6. In the case of the virtual trefoil K, two crossings are odd and, hence, we have $J(K) = 2$. This proves that K is non-trivial, non-classical and inequivalent to its mirror image. Similarly, for the virtual knot E the crossings a and b are odd. We have $J(E) = 2$ and, hence, the knot E is also non-trivial, non-classical and inequivalent to its mirror image. Note that for the knot E the invariant is independent of the type of the crossing c.

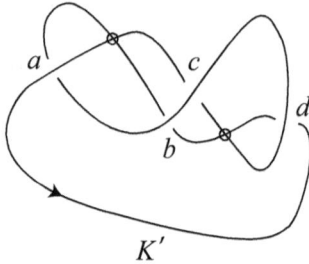

Fig. 1.23   The knot K'

In Fig. 1.23, the virtual knot K' has $J(K') = 2$. Note that K' would be unknotted if we allowed the second (Fig. 1.15, lower part) forbidden move. This example underlines why we forbid such moves in virtual knot theory.

Chapter 2

Parity: definitions and examples

2.1 Introduction

The present chapter and Chap. 4 are devoted to parity and invariants constructed with help of parity.

Let us consider one example of parity. A chord of a chord diagram is said to be *even* if the number of other chords linked to it is even, and *odd* otherwise (we assume that a chord is not linked to itself). It turns out that by using such obvious classification of chords as even and odd we can obtain many non-trivial results in different knot theories.

2.2 Category of knot diagrams

Let $\mathcal{K}$ be a knot. We shall use the notion of 'knot' in one of the following situations:

1) a free knot;
2) a homotopy class of curves immersed in a given surface;
3) a flat knot;
4) a virtual knot.

Let us define the category $\mathfrak{K}$ of diagrams of the knot $\mathcal{K}$. The objects of $\mathfrak{K}$ are diagrams of $\mathcal{K}$ and morphisms of the category $\mathfrak{K}$ are (formal) compositions of *elementary morphisms*. By an *elementary morphism* we mean

27

- an isotopy of diagram;
- a Reidemeister move.

Definition 2.1. A *partial bijection* of sets X and Y is a triple $(\widetilde{X}, \widetilde{Y}, \phi)$, where $\widetilde{X} \subset X$, $\widetilde{Y} \subset Y$ and $\phi \colon \widetilde{X} \to \widetilde{Y}$ is a bijection.

Remark 2.1. Since the number of vertices of a diagram may change under Reidemeister moves, there is no bijection between the sets of vertices of two diagrams connected by a sequence of Reidemeister moves. To construct any connection between two sets of vertices we have introduced the notion of a partial bijection which means just the bijection between the subsets of vertices corresponding to each other in the two diagrams.

Let us denote by $\mathcal{V}$ the *vertex functor* on $\mathfrak{K}$, i.e. a functor from $\mathfrak{K}$ to the category, objects of which are finite sets and morphisms are partial bijections. For each diagram K we define $\mathcal{V}(K)$ to be the set of classical crossings of K, i.e. the vertices of the underlying framed 4-graph. Any elementary morphism $f \colon K \to K'$ naturally induces a partial bijection $f_* \colon \mathcal{V}(K) \to \mathcal{V}(K')$.

2.3 A parity

We are going to generalise the notion of parity for the case of arbitrary abelian group. In [47, 48, 50, 53] the parity with coefficients in $\mathbb{Z}_2$ was defined. We extend that notion to the case with an abelian group. Note that one can define a parity with a non-abelian group, see, for example [54].

Let A be an abelian group.

Definition 2.2. A *parity p on diagrams of a knot $\mathcal{K}$ with coefficients in A* is a family of maps $p_K \colon \mathcal{V}(K) \to A$, $K \in \mathrm{ob}(\mathfrak{K})$, such that for any elementary morphism $f \colon K \to K'$ the following holds:

1) $p_{K'}(f_*(v)) = p_K(v)$ provided that $v \in \mathcal{V}(K)$ and there exists $f_*(v) \in \mathcal{V}(K')$;
2) $p_K(v_1) + p_K(v_2) = 0$ if f is a decreasing second Reidemeister move and v_1, v_2 are the disappearing crossings;
3) $p_K(v_1) + p_K(v_2) + p_K(v_3) = 0$ if f is a third Reidemeister move and v_1, v_2, v_3 are the crossings participating in this move.

Remark 2.2. Note that each knot may have its own group A, and, therefore, different knots generally have different parities.

Lemma 2.1. *Let p be any parity and K be a diagram. Then $p_K(v) = 0$ if f is a decreasing first Reidemeister move applied to K and v is the disappearing crossing of K.*

Proof. Let us apply the second Reidemeister move g to the diagram K as is shown in Fig. 2.1. We have

$$p_{K'}(v_1) + p_{K'}(v_2) = 0, \qquad p_{K'}(g_*(v)) + p_{K'}(v_1) + p_{K'}(v_2) = p_K(v) = 0.$$

$\square$

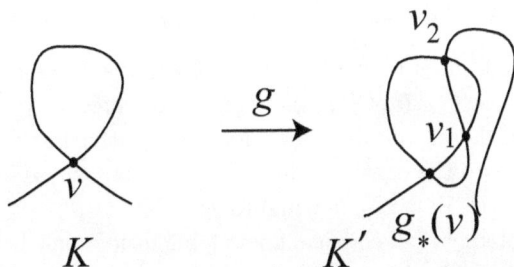

Fig. 2.1   Reduction of the first Reidemeister move to the second and third Reidemeister moves

Let us consider some examples of parities for some knot theories.

2.3.1 *Gaussian parity for free, flat and virtual knots*

Let $A = \mathbb{Z}_2$ and K be a virtual (flat) knot diagram (resp., a framed 4-graph with one unicursal component).

Construct the map $gp_K \colon \mathcal{V}(K) \to \mathbb{Z}_2$ by putting $gp_K(v) = 0$ if the number of vertices linked with v is even (*an even crossing*), and $gp_K(v) = 1$ otherwise (*an odd crossing*).

Statement 2.1 (see [50]). *The map gp is a parity for free, flat and virtual knots.*

Definition 2.3. The parity gp is called the *Gaussian parity*.

2.3.2 Two-component classical and virtual links

Here we call the theory of two-component virtual links just knot theory.

Let $A = \mathbb{Z}_2$ and $L = L_1 \cup L_2$ be a virtual (classical) link diagram.

Define the map $p_L : \mathcal{V}(L) \to \mathbb{Z}_2$ by putting $p_L(v) = 0$ if v is a classical crossing formed by a single component, and $p_L(v) = 1$ if v is a classical crossing formed by two components.

It is not difficult to prove the following.

Statement 2.2. *The map p is a parity.*

2.3.3 Knots in the solid torus, curves on 2-surfaces

Let us consider a class of *knots* in the solid torus B represented as the thickened annulus $S^1 \times I^1 \times I^1$. Knots are represented by their projections on $S^1 \times I^1$ which are obtained by "forgetting" the last factor I^1. We restrict ourselves to the consideration of those knots whose homology class in $H_1(S^1 \times I^1 \times I^1, \mathbb{Z}_2) = \mathbb{Z}_2$ is equal to zero.

On the set of knots (all classical knots lying inside some ball $D^3 \subset B$ are also attributed to this class), we define parity of crossings in the following manner.

Let K be a diagram, and let v be its crossing. Let us smooth the diagram K at the crossing v with respect to the orientation. We obtain a link L. The following equality of homology classes $[K] = [L] = [L_1 \cup L_2] = [L_1 + L_2] \in H_1(S^1 \times I^1 \times I^1, \mathbb{Z}_2) = \mathbb{Z}_2$ is evident. Thus, taking into consideration the equality $[K] = 0$, we get $[L_1] = [L_2] \in \mathbb{Z}_2$.

Let $A = \mathbb{Z}_2$, and let us construct the map $p_K : \mathcal{V}(K) \to \mathbb{Z}_2$ by putting $p_K(v) = 0$ if $[L_1] = [L_2] = 1 \in \mathbb{Z}_2$, and $p_L(v) = 1$, otherwise.

It is not difficult to prove the following.

Statement 2.3. *The map p is a parity.*

Remark 2.3. As a particular case of the parity described above we may consider the parity for theory of closed braids consisting of even number of strands up to an isotopy of braids and up to conjugation.

Each braid can be, in a natural way, represented by a *tangle*, i.e., a diagram inside $I^1 \times I^1$, and a closed braid can be represented as a diagram in $S^1 \times I^1$.

A method using $\mathbb{Z}_2$–homology, can be used in a more general setup.

Virtual knots represent knots in thickened surfaces $S_g \times I$ considered up to isotopy and stabilisation. Let us forget about stabilisation and consider two classes of objects: curves in S_g up to homotopy and knots in $S_g \times I$ up to isotopy. In both cases objects can be encoded by diagrams in general position which represent framed 4-graphs on S_g (in the case of knots, of course, the structure of over/undrcrossing and a cyclic order specified at each crossing). The equivalence relation is defined be means of the Reidemeister moves.

Let us fix a homology class α from $H^1(S_g, \mathbb{Z}_2) = H^1(S_g \times I, \mathbb{Z}_2)$. We shall consider only knots (curves) γ such that $\alpha(\gamma) = 0$.

Then for such set of knots (curves) and for their diagrams on S_g we can define the parity of a crossing by smoothing with respect to the orientation and taking the class α of one of the "halves" L_1 or L_2 obtained after this smoothing.

It is easy to check that this map is a parity.

2.4 Parity and homology

A natural source of parities comes from one-dimensional $\mathbb{Z}_2$–(co)homology classes of the underlying surface of a (virtual) knot. We shall see that if we consider curves in a given closed 2-surface then (modulo some restrictions) these homology classes will lead to well-defined parities for knots on such surfaces (the same works for virtual knots in the thickening of this surface). The inverse statement is also true: if we take a given parity on a given surface, then it will lead to a certain $\mathbb{Z}_2$–homology class of the surface.

So, when we have a knot and a fixed surface associated with it, this gives us a universal receipt of constructing parities and leads us to the universal parity, see ahead.

However, when passing to virtual knots by means of the stabilisation, this causes the following trouble: the surface is not fixed any more and there is no canonical coordinate system on this surface. Thus, for example, if we work on a concrete torus, we may fix a coordinate system on it and take the parity corresponding to the 'meridian'. However, when we stabilise and destabilise, we may destroy the coordinate system on the surface, so it will be impossible to recover the initial (co)homology class.

To this end, we introduce the notion of a characteristic class for underlying surfaces corresponding to virtual knots (see rigorous definition ahead). This is a class which does not depend on anything except a given virtual knot and behaves nicely on surfaces coming from diagrams, in particular, under stabilisations/destabilisations.

We give some concrete examples of constructing characteristic classes.

As we shall see later, this approach does not always help: for the flat knot diagram (in Fig. 2.2) on the surface of genus 2 (the surface is represented as a decagon with opposite sides identified) is so symmetric, that every characteristic class of it is trivial (see Example 2.1), though when we restrict ourselves to this concrete surface of genus 2, there will be non-trivial parities which have non-zero values on the crossings of the flat knot diagram.

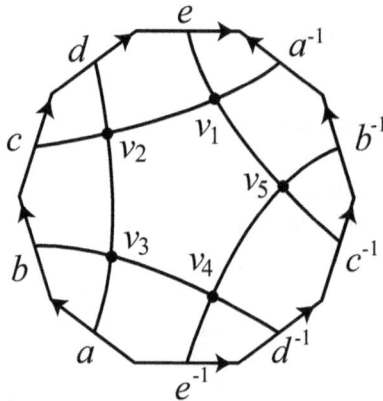

Fig. 2.2   A knot in a surface of genus two

To overcome this difficulty, we enlarge the notion of parity. Instead of a parity valued in $\mathbb{Z}_2$, we introduce the universal parity valued in some linear space over $\mathbb{Z}_2$ which is closely related to knot diagrams (the $\mathbb{Z}_2$–homology group of the underlying space with a fixed basis) and see that all previously known $\mathbb{Z}_2$-valued parities factor through this universal parity.

This parity allows one to work with examples where characteristic classes and their corresponding parities fail.

First of all we describe a connection between a parity and the homologies of a surface.

2.4.1   Homological parity for homotopy classes of curves generically immersed in a surface

Let S be a connected closed surface. We consider a free homotopy class $\mathcal{K}$ of curves generically immersed in S.

Let $A = H_1(S, \mathbb{Z}_2)/[\mathcal{K}]$, where $[\cdot]$ denotes a homological class.

Let K be a framed 4-graph embedded in S representing a curve from $\mathcal{K}$. For each vertex v we have two halves of the graph, $K_{v,1}$ and $K_{v,2}$, obtained by smoothing at this vertex, see Fig. 2.3.

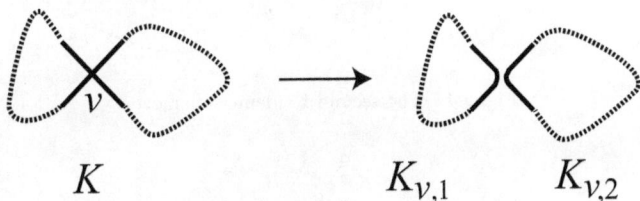

$$K \qquad K_{v,1} \qquad K_{v,2}$$

Fig. 2.3   The graphs $K_{v,1}$ and $K_{v,2}$

Define the map $hp_K \colon \mathcal{V}(K) \to A$ by putting $hp_K(v) = [K_{v,1}]$.

Lemma 2.2 (see [50]). *The map hp is a parity for homotopy classes of curves generically immersed in S.*

Proof. From the definition of A it follows that hp does not depend on the choice of a half for a vertex.

Let $f \colon K \to K'$ be an elementary morphism.

1) Since Reidemeister moves are performed in a small area of S homeomorphic to a disc, we have $hp_{K'}(f_*(v)) = hp_K(v)$ provided that $v \in \mathcal{V}(K)$ and there exists $f_*(v) \in \mathcal{V}(K')$.

2) Let f be a decreasing second Reidemeister move, and let v_1, v_2 be the disappearing crossings. Denote by $K_{v_1,1}$ and $K_{v_2,1}$ the two halves corresponding to the vertices v_1 and v_2, see Fig. 2.4.

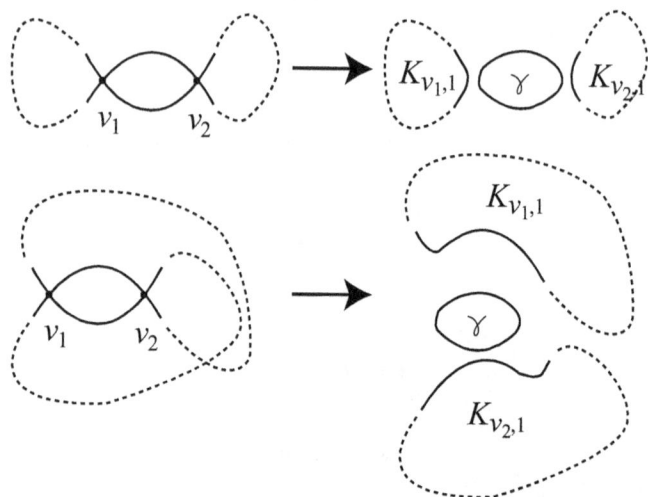

Fig. 2.4   The second Reidemeister move

We have

$$hp_K(v_1) + hp_K(v_2) = [K_{v_1,1}] + [K_{v_2,1}] = [K_{v_1,1}] + [K_{v_2,1}] + [\gamma] = [K] = 0.$$

3) Let f be a third Reidemeister move, and let v_1, v_2, v_3 be the crossings participating in this move. Denote by $K_{v_1,1}$, $K_{v_2,1}$ and $K_{v_3,1}$ the three halves corresponding to v_1, v_2 and v_3 respectively, see Fig. 2.5 (we consider only one case depicted in Fig. 2.5, all other versions of the third Reidemeister move can be treated in the same way).

We have

$$hp_K(v_1) + hp_K(v_2) + hp_K(v_3) = [K_{v_1,1}] + [K_{v_2,1}] + [K_{v_3,1}]$$

$$= [K_{v_1,1}] + [K_{v_2,1}] + [K_{v_3,1}] + [\gamma] = [K] = 0.$$

$\square$

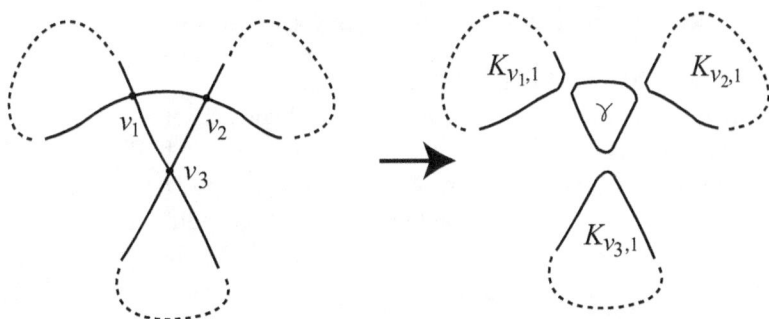

Fig. 2.5 The third Reidemeister move

2.4.2 *Characteristic classes for framed 4-graphs*

Our next task is to understand the topological nature of parity. As we shall see, when we deal with curves on a fixed surface, all possible parities for such curves are closely connected with (co)homology classes with coefficients in $\mathbb{Z}_2$.

However, when we deal with virtual knots or knots in an abstract thickened surface, then there is no canonical choice of the coordinate system on the surface, so we can not say what is a 'cohomology class dual to the longitude' or a 'cohomology class dual to the meridian'. Moreover, cohomology classes have to be chosen in a way compatible with stabilisations.

There is a partial remedy which deals with so-called characteristic classes. Roughly speaking, a characteristic class is a class on the surface corresponding to a knot diagram which can be recovered from the diagram itself. This will be discussed in Sec 2.4.3.

Consider a framed 4-graph K with one unicursal component. The homology group $H_1(K, \mathbb{Z}_2)$ is generated by halves corresponding to vertices. If the set of framed 4-graphs (possibly, with some further decorations at vertices) is endowed with a *parity*, then we can construct the following cohomology class h: for each of the halves $K_{v,1}$, $K_{v,2}$ we set $h(K_{v,1}) = h(K_{v,2}) = p_K(v)$, where $p_K(v)$ is the parity of the vertex v. Taking into account that every two halves for each vertex sum up to give the cycle generated by the whole graph, we have defined a "characteristic" cohomology class h from $H_1(K, \mathbb{Z}_2)$.

Collecting the properties of this cohomology class we see that

(1) For every framed 4-graph K we have $h(K) = 0$.
(2) Let K' be obtained from K by a second Reidemeister move increasing the number of crossings by two. Then for every basis $\{\alpha_i\}$ of $H_1(K, \mathbb{Z}_2)$ there exists a basis in $H_1(K', \mathbb{Z}_2)$ consisting of one "bigon" γ, the elements α_i' naturally corresponding to α_i and one additional element δ, see Fig. 2.6, left.

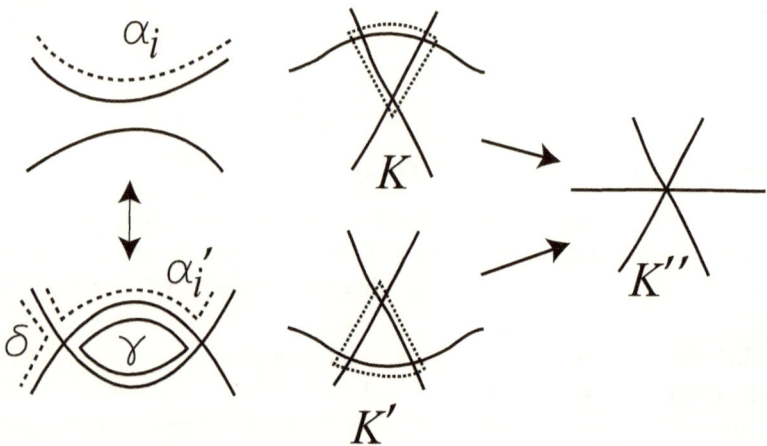

Fig. 2.6   The cohomology condition for Reidemeister moves

Then the following holds: $h(\alpha_i) = h(\alpha_i')$, $h(\gamma) = 0$.

(3) Let K' be obtained from K by a third Reidemeister move. Then there exists a graph K'' with one vertex of valency 6 and the other vertices of valency 4 which is obtained from either of K or K' by contracting the "small" triangle to the point. This generates the mappings $i \colon H_1(K, \mathbb{Z}_2) \to H_1(K'', \mathbb{Z}_2)$ and $i' \colon H_1(K', \mathbb{Z}_2) \to H_1(K'', \mathbb{Z}_2)$, see Fig. 2.6, right.

We require the following to hold: the cocycle h is equal to zero for small triangles, besides that if for $a \in H_1(K, \mathbb{Z}_2)$, $a' \in H_1(K', \mathbb{Z}_2)$ we have $i(a) = i'(a')$, then $h(a) = h(a')$.

Note that in 2 no restriction on $h(\delta)$ is imposed.

Thus, every parity for free knots generates some $\mathbb{Z}_2$–cohomology class for all framed 4-graphs with one unicursal component, and this class behaves nicely under the Reidemeister moves.

The converse is true as well. Assume we are given a certain "universal" $\mathbb{Z}_2$–cohomology class for all framed 4-graphs satisfying the conditions 1–3 described above (later we shall describe the exact definition of the universality). Then it originates from some *parity*. Indeed, it is sufficient to define the parity of every vertex to be the parity of the corresponding half. The choice of a particular half does not matter, since the value of the cohomology class on the whole graph is zero. One can easily check that parity axioms follow.

This point of view allows one to find parities for those knots lying in $\mathbb{Z}_2$–homologically non-trivial manifolds. For more details, see [40].

2.4.3 *Characteristic parities for virtual knots*

Let K be a virtual knot diagram, and let $P = (S, K)$ be the CLSD associated with the diagram K. A *checkerboard colouring* of S with respect to K is a colouring of all the components of $S \setminus K'$, where K' is the image of the embedding of K, by two colours, say black and white, such that two components of $S \setminus K'$ being adjacent by an edge of K' have always distinct colours.

We say that a virtual diagram *admits a checkerboard colouring* or it is *checkerboard colourable* if the associated CLSD admits a checkerboard colouring.

Theorem 2.1 (see [25]). *If two two virtual diagrams admitting a checkerboard colouring are equivalent in the category of virtual knots, then they are equivalent in the category of virtual knots admitting a checkerboard colouring.*

We consider the category of virtual knots admitting a checkerboard colouring.

Definition 2.4. A *characteristic class* of a knot $\mathcal{K} = \{K\}$ is a homology class of the surface S associated with a diagram K such that this class does depend only on $\mathcal{K}$ and behaves nicely under the Reidemeister moves.

Consider the group $H_1(S, \mathbb{Z}_2)$ and any element $[\gamma] \in H_1(S, \mathbb{Z}_2)$. We know that $[K'] = 0$.

Define the map $\chi_{K,\gamma} \colon \mathcal{V}(K) \to \mathbb{Z}_2$ by putting $\chi_{K,\gamma}(v)$ to be equal to the intersection number of γ and $K'_{v,1}$, where $K'_{v,1}$ is a half of K' corresponding to v.

Our aim is to construct a homology class of γ, which does only depend on a virtual knot generated by K, and defines a parity on the virtual knot.

Consider the following cases.

1) Let γ_a be the sum of halves over all classical crossings (for each classical crossing we take only one half).

2) Let $\mathcal{L}$ be an arbitrary non-trivial free link with two linked components. At each vertex of K we can consider a smoothing giving the link diagram with two components. We say that a classical crossing v of K *leads to* $\mathcal{L}$ if after a smoothing of it and considering the result just as a framed 4-graph we get a diagram of $\mathcal{L}$. Let us define

$$\gamma_{\mathcal{L}}(K) = \sum_v K'_{v,1},$$

where the sum is taken over all classical crossings giving a diagram of $\mathcal{L}$.

Theorem 2.2. *The maps χ_{K,γ_a} and $\chi_{K,\gamma_{\mathcal{L}}}$ are parities for virtual knots with coefficients in $\mathbb{Z}_2$.*

Proof. We consider only the map $\chi_{K,\gamma_{\mathcal{L}}}$.

Let $f \colon K_1 \to K_2$ be an elementary morphism of two knot diagrams. Consider two CLSD's $P_1 = (S_1, K_1)$ and $P_2 = (S_2, K_2)$ associated with K_1 and K_2, respectively. It is sufficient to consider two cases:

1) If S_1 and S_2 have the same genus, then the virtue of the claim follows from Lemma 2.2.

2) If the genus of S_2 is smaller than the genus of S_1 by 1, then f is a decreasing second Reidemeister move, see Fig. 2.7.

As $\mathcal{L}$ is a free link then the classical crossings v_1 and v_2 participating in the move either simultaneously give the free link $\mathcal{L}$ or do not give it.

Denote by K'_i the image of K_i in S_i. As any half of any classical crossing of K'_1 intersects any half of a classical crossing distinct from v_1 and v_2 either at 0 or precisely two of v_1, v_2 and we can pick halves $K'_{v_1,i}$ and $K'_{v_2,j}$ in such a way that they are homotopic as curves on S_1, we get

$$\chi_{K_1,\gamma_{\mathcal{L}}}(v_1) + \chi_{K_1,\gamma_{\mathcal{L}}}(v_2) = 0,$$

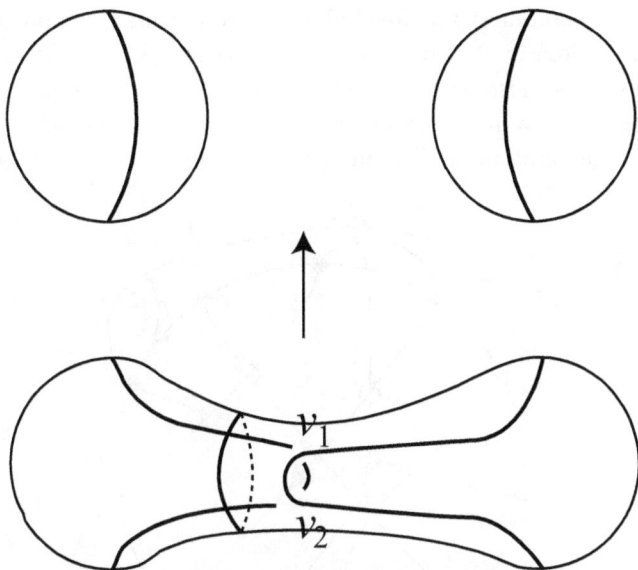

Fig. 2.7 A second Reidemeister move adds a handle

and $\chi_{K_2,\gamma_\mathcal{L}}(f_*(v)) = \chi_{K_1,\gamma_\mathcal{L}}(v)$ provided that $v \in \mathcal{V}(K_1)$ and there exists $f_*(v) \in \mathcal{V}(K_2)$. $\qquad\square$

Example 2.1. Consider the knot diagram K depicted in Fig. 2.2. It is not difficult to show that we have the non-trivial map $hp_K \colon \mathcal{V}(K) \to H_1(S, \mathbb{Z}_2)/[\mathcal{K}]$. The image of this map is the subgroup of $H_1(S, \mathbb{Z}_2)/[\mathcal{K}]$ generated by 5 elements $a_i = hp_K(v_i)$ with the relations $a_1 + a_2 + a_3 + a_4 + a_5 = 0$, cf. [54].

But if we want to construct a characteristic parity with the methods described above we shall fail. K is so symmetric that all five crossings have the same parity, say p. Since we have pentagon, we get $5p = 0$ and, then, $p = 0$.

Let K be an oriented knot diagram. At each classical vertex we have one smoothing respecting the orientation on K. We can construct parity $\chi_{K,\mathcal{L}}$ with an oriented free link $\mathcal{L}$ having two unicursal components by taking the sum only over classical crossings whose smoothings give $\mathcal{L}$.

Let $\mathcal{L}$ be a non-invertible free link with two unicursal components [49], see, for example, Fig. 2.8. If a vertex of an oriented knot leads to $\mathcal{L}$, then this vertex does most probably not lead to $\overline{\mathcal{L}}$, where $\overline{\mathcal{L}}$ is the free link obtained from $\mathcal{L}$ by reversion of the orientation. It means that a parity does feel an orientation on diagrams.

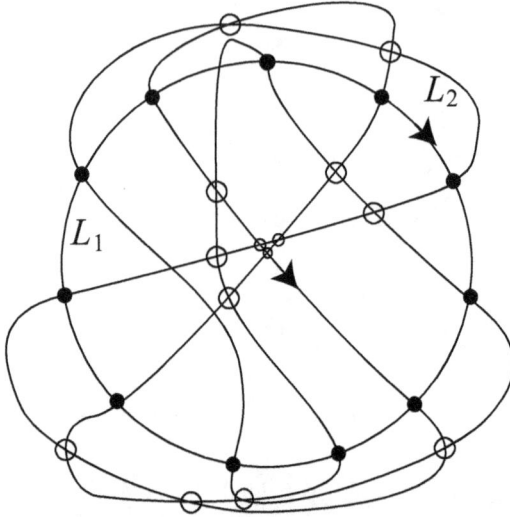

Fig. 2.8   A non-invertible free link

Chapter 3

The universal parity

3.1 Introduction

In Chap. 2 we have given a receipt how to construct parity from homology classes and indicated how to construct characteristic homology classes from the knot itself; these classes lead to concrete parities. However, when we apply such characteristic classes to the knot in Fig. 2.2, we see that all corresponding parities vanish. Nevertheless, the corresponding flat knot lies in the surface S_2 of genus 2 and is not contractible. So, there are some homology classes (which are presumably not characteristic) which yield some parity for some coordinate system of S_2, which is non-trivial on some vertices of the knot. The idea of the present chapter is to construct the universal parity, cf. [54], valued in a certain group related to the knot rather than the group $\mathbb{Z}_2$. This parity will be universal in the sense that any concrete parity on a given surface factors through the universal one.

Definition 3.1. A parity p_u with coefficients in A_u is called a *universal parity* if for any parity p with coefficients in A there exists a unique homomorphism of groups $\rho \colon A_u \to A$ such that $p_K = \rho \circ (p_u)_K$ for any diagram K.

Let us describe a construction of the universal parity in general case.

Let K be a knot diagram. Denote by $1_{K,v}$ the generator of the direct summand in the group $\bigoplus_K \bigoplus_{v \in \mathcal{V}(K)} \mathbb{Z}$ corresponding to the vertex v of K.

Let A_u be the group

$$A_u = \left(\bigoplus_K \bigoplus_{v \in \mathcal{V}(K)} \mathbb{Z} \right) / \mathcal{R},$$

where $\mathcal{R}$ is the set of relations of four types:

1) $1_{K',f_*(v)} = 1_{K,v}$ if $v \in \mathcal{V}(K)$ and there exists $f_*(v) \in \mathcal{V}(K')$;
2) $1_{K,v_1} + 1_{K,v_2} = 0$ if f is a decreasing second Reidemeister move and v_1, v_2 are the disappearing crossings;
3) $1_{K,v_1} + 1_{K,v_2} + 1_{K,v_3} = 0$ if f is a third Reidemeister move and v_1, v_2, v_3 are the crossings participating in this move.

For each diagram K the map $(p_u)_K$ is defined by the formula $(p_u)_K(v) = 1_{K,v}$, $v \in \mathcal{V}(K)$.

If p is a parity with coefficients in a group A, one defines the map $\rho: A_u \to A$ in the following way:

$$\rho \left(\sum_{K,\, v \in \mathcal{V}(K)} \lambda_{K,v} 1_{K,v} \right) = \sum_{K,\, v \in \mathcal{V}(K)} \lambda_{K,v} p_K(v), \quad \lambda_{K,v} \in \mathbb{Z}.$$

The examples below present explicit description of the universal parity.

3.2   Free knots

In the present section we show that in the case of free knot theory there exists only one non-trivial parity, the Gaussian parity.

Theorem 3.1. *Let $\mathcal{K}$ be a free knot. Then the Gaussian parity (with coefficients in $\mathbb{Z}_2$) on diagrams of $\mathcal{K}$ is the universal parity.*

Remark 3.1. Theorem 3.1 means that for each free knot and for each parity on it either all vertices are even or they have the Gaussian parity.

This theorem will follow from Lemmas 3.1, 3.2, 3.3.

We consider free knots as Gauss diagrams with an ordered collection of distinct chords $\{a_1, \ldots, a_n\}$. Let us choose a point distinct from ends of chords on the core circle of a chord diagram. When going around the circle from the chosen point counterclockwise order we shall meet each chord end.

Denoting each end of a chord by the same letter as the chord we shall get a word, where each letter corresponds to a chord and occurs precisely twice.

Definition 3.2. Let D be a chord diagram. We shall say that an ordered collection of chords with numbers $i_1, \ldots, i_k$ of D forms a *polygon*, if a word corresponding to D contains the following sequences of distinct letters $b_{2p-1} b_{2p}$, where $b_{2p-1}, b_{2p} \in \{a_{i_{\sigma(p)}}, a_{i_{\sigma(p-1)}}\}$, $p = 1, \ldots, k$, for some permutation $\sigma \in S_k$.

The pairs (b_{2p-1}, b_{2p}) of letters b_{2p-1}, b_{2p} from the definition of a polygon are said to be *sides* of the polygon.

Example 3.1. Consider the chord diagrams depicted in Fig. 3.1. The chords denoted by a_2, a_4, a_5, a_6, a_8 form a convex pentagon (left) and a non-convex pentagon (right).

In Fig. 3.2 we depict a hexagon for a knot diagram. The knot diagram does not intersect the interior of the hexagon.

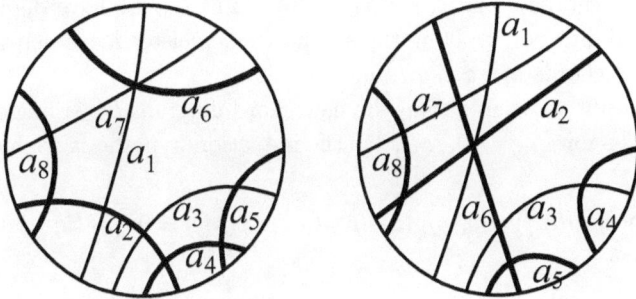

Fig. 3.1 Pentagons

Lemma 3.1. *For every parity and any chord diagram the sum of the parities of chords forming a polygon is equal to* 0.

Remark 3.2. The claim of Lemma 3.1 can be taken as a definition of a parity, see [54].

Proof. Let p be an arbitrary parity on chord diagrams of the free knot $\mathcal{K}$, and let D be a chord diagram representing $\mathcal{K}$. Let us prove the claim of the lemma by induction over the number of sides of polygons.

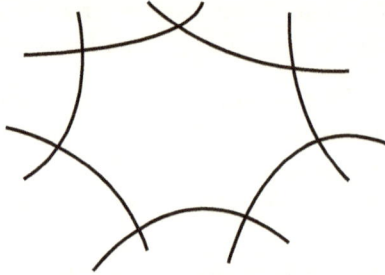

Fig. 3.2 A hexagon

The induction base. The virtue of the claim for a loop, bigon, triangle follows from Lemma 2.1 and Definition 2.2, respectively.

The induction step. Assume that the claim is true for $(k-1)$-gons. Let us consider an arbitrary k-gon $a_{i_1} a_{i_2} \ldots a_{i_k}$.

Let us apply the second Reidemeister move to the chord diagram D by adding two chords b and c, see Fig. 3.3 (in Fig. 3.3 we have depicted the three possibilities of applying the second Reidemeister move depending on the ends of chords a_{i_1}, a_{i_2}, a_{i_3}, a_{i_k}).

As a result we shall obtain the new chord diagram D', triangle $b\,a_{i_1} a_{i_2}$ and $(k-1)$-gon $c\,a_{i_3} a_{i_4} \ldots a_{i_k}$. By the induction hypothesis we have

$$p_{D'}(c) + \sum_{j=3}^{k} p_{D'}(a_{i_j}) = 0, \; p_{D'}(b) + p_{D'}(a_{i_1}) + p(a_{i_2}) = 0, \; p_{D'}(b) + p_{D'}(c) = 0.$$

Therefore,

$$\sum_{j=1}^{k} p_{D'}(a_{i_j}) = \sum_{j=1}^{k} p_D(a_{i_j}) = 0.$$

$\square$

Remark 3.3. If we work with knot diagrams, then the corresponding picture for Lemma 3.1 looks like as is shown in Fig. 3.4.

Let us pass from free knot theory to flat knot theory and virtual knot theory. Since bigons and triangles participating in Reidemeister moves can be spanned by discs we get the following statement.

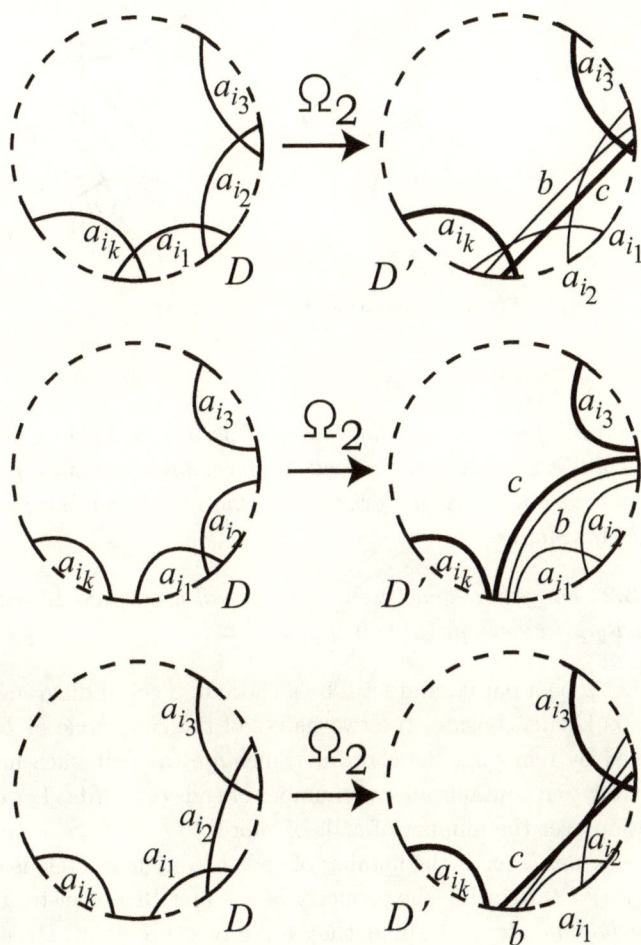

Fig. 3.3 The second Reidemeister move

Corollary 3.1. *For every parity and any flat (virtual) knot diagram the sum of the parities of crossings forming a polygon, which is spanned by a disc in the underlying surface, is equal to* 0.

By using virtualisation moves we can transform any polygon to a polygon which is spanned by a disc in the underlying surface. As a result we

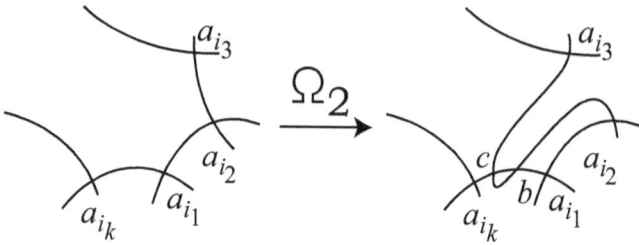

Fig. 3.4   The second Reidemeister move

get the following statement.

Corollary 3.2. *If we consider the theory of pseudo-knots, i.e., the theory of virtual knots modulo the virtualisation move, then Lemma 3.1 remains true in this theory too, i.e., the existence of the writhe number gives us no additional information.*

Lemma 3.2. *For a free knot (pseudo-knot) with a diagram K and an arbitrary parity p we have $p_K(a) = 0$ if $gp_K(a) = 0$.*

Proof. Let p be a parity, and let a be a chord of a chord diagram D with $gp_D(a) = 0$. Let us consider the two halves of the core circle of D, which are obtained by removing the chord a. Since $gp_D(a) = 0$, each half-circle corresponding to a contains an even number of ends of chords. Let us apply the induction over the number of ends of chords.

The induction base: If the number of ends on any half-circle is equal to 0, then $p_D(a) = 0$ by using the property of the first Reidemeister move.

The induction step: Assume that for any chord d of D such that $gp_D(d) = 0$ and a half-circle contains less than $n = 2k$ ends of chords, we have $p_D(d) = 0$. Let us consider a chord a such that one of its half-circles, $K_{a,1}$, contains exactly n ends of chords and the other one, $K_{a,2}$, contains more than or equal to n ends.

Let us orient D in counterclockwise manner and consider the following two cases.

1) The first two ends in $K_{a,1}$ belong to two distinct chords a_1, a_2, see Fig. 3.5. Apply the second increasing Reidemeister move by adding a pair of chords b, b' in such a way that the half-circle corresponding to b' would

contain the set of ends lying in $K_{a,1}$ minus the first ends of a_1, a_2, see Fig. 3.6 (above). Let us show that $p_{D'}(a) + p_{D'}(b) = 0$ in the new chord diagram D'. Let us add the pair of chords c, c' to form the triangle $a_1 a_2 c$, see Fig. 3.6 (below). Then $p_{D''}(a_1) + p_{D''}(a_2) + p_{D''}(c) = 0$ in D''. Moreover, we have the pentagon $a a_1 c a_2 b$ and, therefore, the following equality holds (Lemma 3.1)

$$p_{D''}(a) + p_{D''}(a_1) + p_{D''}(c) + p_{D''}(a_2) + p_{D''}(b) = 0.$$

We get $p_{D''}(a) + p_{D''}(b) = 0$ and $p_{D'}(a) + p_{D'}(b) = 0$. In the half-circle corresponding to b' the number of ends is less than the number of ends in the half-circle corresponding to a. By the induction hypothesis we get $p_{D'}(b) = p_{D'}(b') = 0$, and $p_D(a) = 0$.

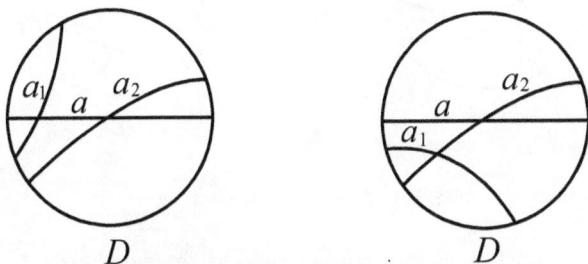

Fig. 3.5   The Gaussian parity zero

2) If the first two ends belong to the same chord c, then $p_D(c) = 0$ (the first Reidemeister move) and c forms the triangle in D' with the chords a and b. Therefore, $p_{D'}(a) + p_{D'}(b) + p_{D'}(c) = 0$. By the induction hypothesis we get $p_{D'}(b) = p_{D'}(b') = 0$ and $p_D(a) = p_{D'}(b) = 0$. $\square$

Lemma 3.3. *Let p be an arbitrary parity (with coefficients from a group A) on diagrams of the free knot represented by a chord diagram D. Then for any two chords a, b such that $gp_D(a) = gp_D(b) = 1$, we have $p_D(a) = p_D(b) = x \in A$ and $2x = 0$.*

Proof. Let $c_1, \ldots, c_k$ be ends of chords lying between the nearest ends of a and b.

Apply k times the second Reidemeister moves as it is shown in Fig. 3.7 (in the centre). Let us show that $p_{D'}(d_l) = (-1)^l x$, where $x = p_{D'}(a)$.

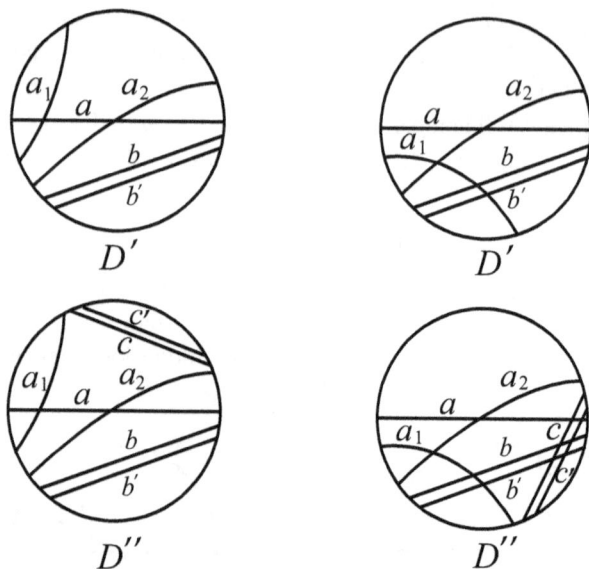

Fig. 3.6   The Gaussian parity zero

Apply the second Reidemeister move by adding two chords f, f' to form the triangle $ad_1 f$. We have

$$gp_{D''}(a) = gp_{D''}(d_1) = 1 \Longrightarrow gp_{D''}(f) = 0 \Longrightarrow p_{D''}(f) = 0$$

$$\Longrightarrow p_{D'}(d_1) = p_{D''}(d_1) = -x.$$

By the induction we can prove that $p_{D'}(d_l) = (-1)^l x$ and $p_D(b) = (-1)^{k+1} x$.

Let us apply the third Reidemeister move to the triangle $ad_1 f$. The parity p and the Gaussian parity of the chord a do not change but the parity of the number of ends of chords between a and b changes. Applying the previous trick we get $p_D(b) = (-1)^k x$, i.e., $2x = 0$.

□

By using Lemmas 3.2 and 3.3 for any parity p (with coefficients from a group A) on diagrams of the free knot having a diagram K we can construct the homomorphism $\rho \colon A \to \mathbb{Z}_2$ by taking $\rho(x) = 1$, where $p_K(a) = x$ and $gp_K(a) = 1$. This concludes the proof of Theorem 3.1.

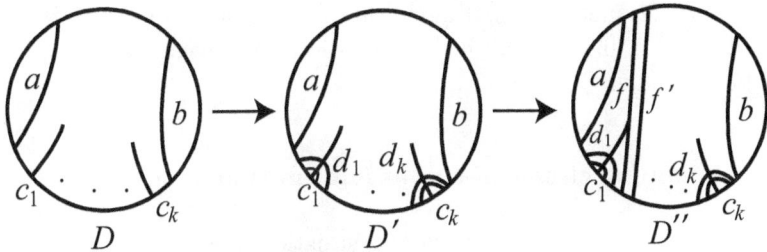

Fig. 3.7 The Gaussian parity one

Remark 3.4. Let p be a parity on a free knot $\mathcal{K}$. It is not possible that there exist two diagrams K_1 and K_2 of $\mathcal{K}$, both having chords being odd in the Gaussian parity such that p is trivial on K_1, and p is the Gaussian parity on K_2. It follows from the fact that there exists a sequence of Reidemeister moves transforming K_1 to K_2 such that any diagram in this sequence has chords being odd in the Gaussian parity.

Before passing to classical knots, we should point out the following. It is known that classical knot theory embeds in virtual knot theory [22, 33]. This means that if two classical knot (link) diagrams are virtually equivalent, then they are isotopic (classically equivalent).

Nevertheless, the parity axiomatic applied to the classical knot theory as a part of the virtual knot theory and to the classical knot theory as it is, should be treated differently.

Namely, from the above we get the following theorem.

Theorem 3.2. *Any parity on virtual knots (one-component knots, not links) is trivial on any classical knots.*

By itself, it does not guarantee that there is no non-trivial parity on classical knots: possibly, there might be some which does not extend to virtual knots? Indeed, for the classical knot theory as it is we are restricted only to those diagrams having classical crossings, and some "additional" crossings used to prove the above lemmas can make the diagram classical.

However, the following theorem holds as well.

Theorem 3.3. *For classical knot theory there exists a unique parity, the trivial parity.*

The proof is indeed a slight modification of Theorem 3.1, which is based on Lemmas 3.2 and 3.3. We just use classical knot diagrams on the plane and bear in mind Corollary 3.1.

3.3 Homotopy classes of curves immersed in a surface

In the previous section we have the situation when all polygons "are spanned" by discs on the plane. Now we are interested in those polygons which are spanned by discs in a surface. Thus, we shall naturally come to the homology of the surface.

Theorem 3.4. *Let $\mathcal{K}$ be a homotopy class of curves generically immersed in a surface S. Then the homological parity (with coefficients in $H_1(S, \mathbb{Z}_2)/[\mathcal{K}]$) is the universal parity on curves of $\mathcal{K}$.*

Proof. We start the proof of the theorem with the following general lemmas.

Lemma 3.4. *Let p be a parity, K be a curve on S and $a \in \mathcal{V}(K)$. Then $2p_K(a) = 0$.*

Proof. By applying the second and third Reidemeister moves we get curves K_1 and K_2 (see Fig. 3.8). We have the equality $p_{K_1}(a) + p_{K_1}(b) = 0$. Then $p_{K_2}(a) + p_{K_2}(b) = 0$. We also have $p_{K_2}(a) + p_{K_2}(c) + p_{K_2}(d) = 0$ and $p_{K_2}(b) + p_{K_2}(c) + p_{K_2}(d) = 0$. Hence, $p_{K_2}(a) = p_{K_2}(b)$ and $2p_{K_2}(a) = 0$. Then $2p_{K_1}(a) = 0$ and $2p_K(a) = 0$. □

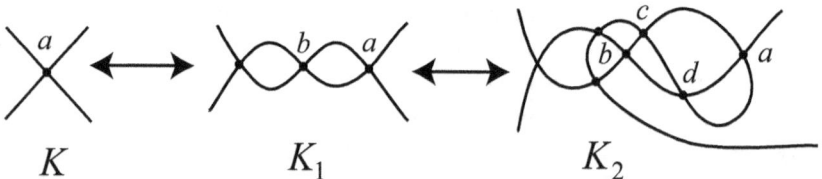

Fig. 3.8 The second and third Reidemeister moves

Lemma 3.5 (cf. [54]). *Let K be a framed 4-graph with one unicursal component. Consider K as a 1-dimensional cell complex. Then $H_1(K, \mathbb{Z}_2)/[K] \cong \bigoplus_{v \in \mathcal{V}(K)} \mathbb{Z}_2$.*

Proof. Let C be the chord diagram corresponding to K. Then C and K are homotopy equivalent as topological spaces. Let C' (resp., K') is the topological space obtained by gluing to C (resp., K) a 2-disc along the core circle of C. Then C' and K' are homotopy equivalent to each other too and $H_1(C', \mathbb{Z}_2) \cong H_1(K', \mathbb{Z}_2) = H_1(K, \mathbb{Z}_2)/[K]$. On the other hand, C' is homotopy equivalent to the bouquet of circles corresponding to the cords of the diagram C, i.e., the crossings of K. Hence, $H_1(C', \mathbb{Z}_2) \cong \bigoplus_{v \in \mathcal{V}(K)} \mathbb{Z}_2$. $\square$

The isomorphism of the lemma identifies the generator of the group $\mathbb{Z}_2$ corresponding to a vertex $v \in \mathcal{V}(K)$ with the homology class $[K_{v,1}] = [K_{v,2}] \in H_1(K, \mathbb{Z}_2)/[K]$.

Lemma 3.6. *Let ω be a closed path on the curve K with rotation points $v_1, v_2, \ldots, v_k$. Then $[\omega] = \sum_{i=1}^{k} [K_{v_i}] \in H_1(K, \mathbb{Z}_2)/[K]$.*

Proof. By attaching a half $K_{v_i,j}$ for each vertex v_i to the path ω we get a closed path without rotation points, i.e., a multiple of K. Thus,

$$[\omega] + \sum_{i=1}^{k} [K_{v_i}] = m[K] = 0.$$

$\square$

Let us return now to the proof of Theorem 3.4.

Let p be a parity with coefficients in a group A on curves of a homotopy class $\mathcal{K}$ on a closed 2-surface S.

Let K be a curve from $\mathcal{K}$ on the surface S. Assume that K splits the surface into a union of 2-cells. Arguing as above in Lemma 3.1, we obtain the following lemma.

Lemma 3.7. *Let e be a cell in $S \backslash K$ with vertices $v_1, \ldots, v_k$ (not necessarily distinct). Then $\sum_{i=1}^{k} p_K(v_i) = 0$.* $\square$

Let us show that the map $\rho_K \colon H_1(S, \mathbb{Z}_2)/[\mathcal{K}] \to A$ given by the formula $\rho([K_{v,1}]) = p_K(v)$, $v \in \mathcal{V}(K)$, is well defined.

The group $H_1(S, \mathbb{Z}_2)/[K]$ is the first homology group of the topological space S' obtained from S by gluing a disc along K. S' can also be considered as the result of gluing cells $e \in S \setminus K$ to the space K' of Lemma 3.5. Hence,

$$H_1(S, \mathbb{Z}_2)/[K] = (H_1(K', \mathbb{Z}_2)/[K]) / ([\partial e], \ e \in S \setminus K)$$

$$= \bigoplus_{v \in \mathcal{V}(K)} \mathbb{Z}_2[K_{v,1}] \Big/ \left(\sum_{v \in e \cap \mathcal{V}(K)} [K_{v,1}] = 0, \ e \in S \setminus K \right)$$

$$= \bigoplus_{v \in \mathcal{V}(K)} \mathbb{Z} 1_{K,v} \Big/ \left(2 \cdot 1_{K,v} = 0, v \in \mathcal{V}(K); \ \sum_{v \in e \cap \mathcal{V}(K)} 1_{K,v} = 0, e \in S \setminus K \right).$$

The second equality follows from Lemmas 3.5 and 3.6.

On the other hand, due to Lemmas 3.4 and 3.7 we have identities $2p_K(v) = 0$, $v \in \mathcal{V}(K)$, and $\sum_{v \in e \cap \mathcal{V}(K)} p_K(v) = 0$, $e \in S \setminus K$, which imply that the map ρ is well defined epimorphism of groups.

Let $f \colon K \to K'$ be an elementary morphism (an isotopy or a Reidemeister move) and the diagram K' splits the surface into cells. Then for any vertex $v' \in \mathcal{V}(K')$ such that $v' = f_*(v)$ for some $v \in \mathcal{V}(K)$ we have $[K_{v,1}] = [K'_{v',1}]$ and $p_K(v) = p_{K'}(v')$. Since the elements $[K'_{v',1}]$ for such vertices v' generate the group $H_1(S, \mathbb{Z}_2)/[\mathcal{K}]$ the maps ρ_K and $\rho_{K'}$ coincide. Hence, the map $\rho = \rho_K$ does not depend on a choice of the diagram K and $p_K = \rho \circ hp_K$ for any diagram which splits the surface into cells.

If $S \setminus K$ is not a union of cells, then we can apply second Reidemeister moves several times and obtain a diagram K' splitting the surface into cells. By properties of the parities hp and p we have $[K_{v,1}] = [K'_{f_*(v),1}]$ and $p_K(v) = p_{K'}(f_*(v))$ for any $v \in \mathcal{V}(K)$. Therefore $p_K(v) = p_{K'}(f_*(v)) = \rho \circ hp_{K'}(f_*(v)) = \rho \circ hp_K(v)$.

Thus, $p_K = \rho \circ hp_K$ for any diagram K, so the homological parity hp is universal. $\qquad\square$

The homological parity remains universal if we pass from the category of homotopy classes of curves on a given surface S to the category of knots on S (to be precise, knots in the thickened surface). The following lemma shows that in some sense parity does not feel the over- and undercrossing structure.

Lemma 3.8. *Let p be a parity on the category of knots on a surface S, and let K be a diagram of a knot on S. If vertices a, b ∈ $\mathcal{V}(K)$ form a bigon in S, then $p_K(a) + p_K(b) = 0$. If vertices v_1, v_2, $v_3 \in \mathcal{V}(K)$ form a triangle in S, then $p_K(a) + p_K(b) + p_K(c) = 0$.*

Proof. We prove the lemma for a triangle, the proof for a bigon is analogous. Let the vertices a, b, c ∈ $\mathcal{V}(K)$ form a triangle. If one can apply the third Reidemeister move to the triangle, the identity $p_K(a) + p_K(b) + p_K(c) = 0$ follows from definition of parity. Otherwise the vertices constitute an alternating triangle. By applying three second and one third Reidemeister moves we get the diagram K' (see Fig. 3.9), where the following equalities hold:

$$p_{K'}(b) + p_{K'}(c) + p_{K'}(d) = 0,$$
$$p_{K'}(e) + p_{K'}(f) + p_{K'}(g) = 0,$$
$$p_{K'}(a) + p_{K'}(f) + p_{K'}(g) = 0,$$
$$p_{K'}(e) + p_{K'}(d) = 0.$$

Then we have $p_{K'}(a) = p_{K'}(e) = p_{K'}(d) = p_{K'}(b) + p_{K'}(c)$ (we do not need signs because Lemma 3.4 remains true in the category of knots). Therefore, $p_K(a) + p_K(b) + p_K(c) = 0$.   □

The claim above ensures that Lemma 3.7 holds in the current situation too. Hence, one can repeat the proof of Theorem 3.4 and get the following result.

Theorem 3.5. *Let $\mathcal{K}$ be a knot on a surface S. Then the homological parity (with coefficients in $H_1(S, \mathbb{Z}_2)/[\mathcal{K}]$) is the universal parity on diagrams of $\mathcal{K}$.*

Corollary 3.3. *Any parity on classical knots is trivial.*

Proof. Any classical knot $\mathcal{K}$ is represented by diagrams on S^2. But $H_1(S^2, \mathbb{Z}_2) = 0$, so the universal parity group as well as any parity is trivial.   □

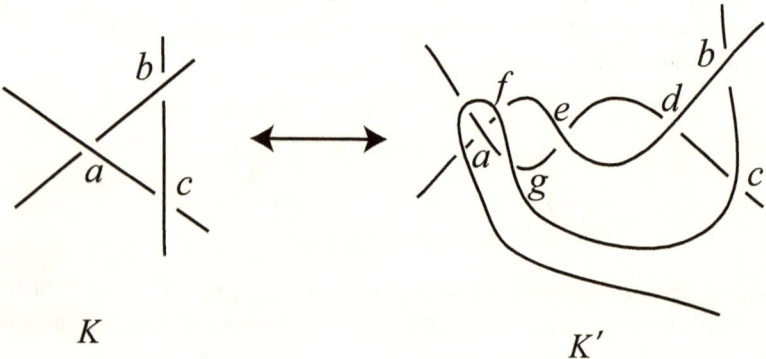

Fig. 3.9   An alternating triangle

Chapter 4

Applications of parity

4.1 Introduction

In this chapter we use a parity for constructing invariants of knots.

The first part of the chapter is devoted to functorial maps (weak parities). This map can be used in order to "lift" all invariants of some knot theory to the realm of knots in another theory. For example, many invariants of classical knots do not admit any evident generalisation for the case of virtual knots: in some cases, e.g. for Khovanov homology, one has to revisit completely the original definition in the classical case, whence some other invariants rely on geometry and topology of the 3-space. The presence of a "right" projection usually allows one not only to lift many invariants, but also to refine them in many different ways. Then by using parities we construct various types of brackets. For constructing these brackets we shall use approaches developed by Goldman, Turaev and Kauffman. Also in this chapter we construct relative parities and parities obtained from brackets.

4.2 Functorial maps and weak parities

4.2.1 *Functorial maps*

Definition 4.1. Let $\mathcal{K}$ be a knot. A *functorial map* on the diagram category $\mathfrak{K}$ of the knot $\mathcal{K}$ is a functor Ψ from the category $\mathfrak{K}$ to the category of virtual knot diagrams (if $\mathcal{K}$ is a virtual knot or a knot in a given surface) or to the category of flat diagrams (if $\mathcal{K}$ is a flat knot or a free homotopy class of a curve in a given surface) such that

1) for any diagram $D \in \text{ob}(\mathfrak{K})$ the diagram $\Psi(D)$ is obtained from D by replacement of some classical crossings by virtual ones;
2) for any Reidemeister move $f \colon D_1 \to D_2$ the map
$$\Psi(f) \colon \Psi(D_1) \to \Psi(D_2)$$
is *the same* Reidemeister move or a detour move.

Let us consider a diagram $D \in \text{ob}(\mathfrak{K})$. Let $\mathcal{V}(D)$ be the set of classical crossings of the diagram. Any elementary morphism $f \colon D \to D'$ (i.e., isotopy, Reidemeister move or a detour move) induces a correspondence between the classical crossings of the diagrams, that is some partial map $f_* \colon \mathcal{V}(D) \to \mathcal{V}(D')$. If the map f is an isotopy, detour move or a third Reidemeister move, then the domain of definition (resp., image) of f_* is the whole set $\mathcal{V}(D)$ (resp., $\mathcal{V}(D')$), and if f is a first or second Reidemeister move, then the domain of definition (resp., image) is the set of crossing of the diagram D (resp., D') which are not involved in the move. The map f_* is a bijection of the domain of definition onto the image.

Let Ψ be a functorial map. Then it induces on the set of classical crossings $\mathcal{V}(D)$ of any diagram $D \in \text{ob}(\mathfrak{K})$ a map $\psi_D \colon \mathcal{V}(D) \to \mathbb{Z}_2$ such that $\psi_D(v) = 0$ if the crossing v remains in the diagram $\Psi(D)$, and $\psi_D(v) = 1$ if the crossing v becomes virtual in the diagram $\Psi(D)$. We shall call the crossing v to be *even* with respect to the functorial map Ψ if $\psi_D(v) = 0$, and *odd* if $\psi_D(v) = 1$.

Proposition 4.1. *The maps ψ_D have the following properties:*
($\Psi 0$) *for any Reidemeister move $f \colon D \to D'$ and any crossing $v \in \mathcal{V}(D)$, which is not involved in the move, we have $\psi_D(v) = \psi_{D'}(f_*(v))$;*
($\Psi 2$) *if $f \colon D \to D'$ is a decreasing second Reidemeister move on crossings $u, v \in \mathcal{V}(D)$, then $\psi_D(u) = \psi_D(v)$;*
($\Psi 3$) *if $f \colon D \to D'$ is a third Reidemeister move on crossings $u, v, w \in \mathcal{V}(D)$, then $\psi_D(u) = \psi_{D'}(f_*(u))$, $\psi_D(v) = \psi_{D'}(f_*(v))$, $\psi_D(w) = \psi_{D'}(f_*(w))$ and $\psi_D(u) + \psi_D(v) + \psi_D(w) \neq 1$ (in $\mathbb{Z}$).*

Proof. Let a crossing v be not involved in the move. Then the corresponding (classical or virtual) crossing $\Psi(v)$ in the diagram $\Psi(D)$ is not involved in the move $\Psi(f)$. Hence, the crossing $\Psi(v)$ and its image $\Psi(f_*(v))$

in the diagram $\Psi(D')$ have the same type (classical or virtual). Thus, $\psi_D(v) = \psi_{D'}(f_*(v))$.

Let f be a third Reidemeister move on crossings u, v, w. In order to obtain a third move on the same vertices or a detour move, one should make virtual $0, 2$ or 3 crossings among u, v, w; replacement of one crossing to virtual crossing leads to a forbidden move (see Fig. 4.1). Moreover, the parities of the corresponding crossings in the diagrams $\Psi(D)$ and $\Psi(D')$ must coincide. Thus, the property ($\Psi 3$) holds.

The proof for the property ($\Psi 2$) is analogous. $\qquad\square$

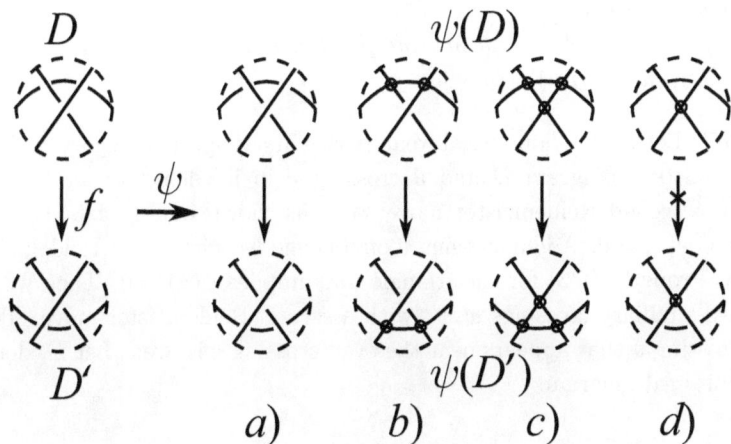

Fig. 4.1 Proof for ($\Psi 3$): a) when all the crossings remain classical one has a third Reidemeister move; b), c) when two or three crossings are replaced with virtual crossings a detour move appears; d) replacement of one crossing to virtual leads to a forbidden move

Definition 4.2. Any family of maps $\psi_D \colon \mathcal{V}(D) \to \mathbb{Z}_2$, $D \in \mathrm{ob}(\mathfrak{K})$, that possesses the properties ($\Psi 0$), ($\Psi 2$), ($\Psi 3$) is called a *weak parity*.

Note that the correspondence between functorial maps and weak parities established above is one-to-one, i.e., the notions of functorial map and weak parity are equivalent. Below we shall exploit mainly the notion of weak parity.

Example 4.1. Let us define a weak parity $\mathbb{O}$ with the identity $\mathbb{O}_D(v) \equiv 0$. This parity is called a *null weak parity*. The null parity corresponds to the identity functorial map.

Example 4.2. A *trivial weak parity* $\mathbb{1}$ is defined by equality $\mathbb{1}_D(v) \equiv 1$. It corresponds to the functorial map which makes all the classical crossings virtual. So the map produces pure virtual diagrams of the unknot.

Below we shall consider only non-trivial weak parities.

Theorem 4.1. *Let a weak parity ψ be non-trivial. Then it possesses the following property*:
(Ψ1) *if $f \colon D \to D'$ is a decreasing fist Reidemeister move on a crossing $u \in \mathcal{V}(D)$, then $\psi_D(u) = 0$.*

Proof. Let ψ be a non-trivial weak parity in a diagram category $\mathfrak{K}$. Then there exists a diagram D and a crossing v in it such that $\psi_D(v) = 0$. Apply a second Reidemeister move to D as shown in Fig. 4.2. By (Ψ2) $\psi(w) = \psi(v) = 0$. Add a crossing u on an edge vw with a first Reidemeister move. Property (Ψ3) for the triangle uvw implies $\psi(u) = 0$. Then we can remove auxiliary crossings w, w' with a second Reidemeister move. Thus, the crossing u that appears near the even crossing v is even. Let D' denote the obtained diagram.

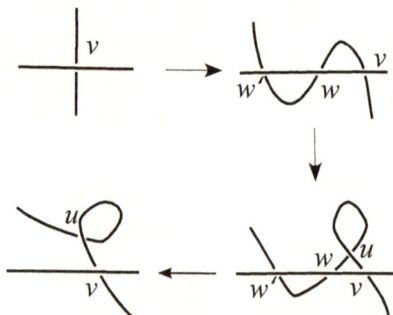

Fig. 4.2 Proof of property (Ψ1)

Let $f: D_1 \to D_2$ be an increasing first Reidemeister move in the diagram category $\mathfrak{K}$ and $t \in \mathcal{V}(D_2)$ be the new crossing. Since diagrams D and D_1 are diagrams of the same knot, there is a sequence of Reidemeister moves which transforms D into D_1. Then we can construct a sequence of moves (inserting second and third Reidemeister moves if necessary) which transforms D' into D_2 and maps the crossing u into the crossing t. By properties ($\Psi 0$) and ($\Psi 3$) we have $\psi_{D_2}(t) = \psi_{D'}(u) = 0$. $\square$

A number of non-trivial weak parities can be obtained from ordinary parities.

Let G be a group. Assume that a knot $\mathcal{K}$ is oriented, so on the diagrams of the diagram category $\mathfrak{K}$ the induced orientation is given.

Definition 4.3. An *oriented parity* p on the diagram category $\mathfrak{K}$ is a family of maps $p_D: \mathcal{V}(D) \to G$, $D \in \mathrm{ob}(\mathfrak{K})$, that possesses the following properties:

- for any elementary morphism $f: D \to D'$ and any crossing $v \in \mathcal{V}(D)$ in the domain of the partial map f_* one has $p_D(v) = p_{D'}(f_*(v))$;
- if $f: D \to D'$ is a decreasing first Reidemeister move and $v \in \mathcal{V}(D)$ is the disappearing crossing, then $p_D(v) = 1$ where $1 \in G$ is the unit of the group;
- if $f: D \to D'$ is a decreasing second Reidemeister move or a third Reidemeister move then $\prod p_D(v_i)^{\epsilon(v_i)} = 1$, where v_i are the vertices involved in the move and $\epsilon(v_i)$ is the incidence index of the vertex v_i in relation with the face that takes part in the move (bigon or triangle), see Fig. 4.3 (left), and the order of the vertices in the product is induced by the orientation of the surface the diagram lies in.

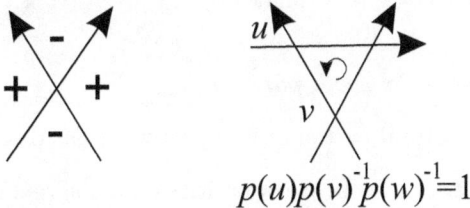

$$p(u)p(v)^{-1}p(w)^{-1} = 1$$

Fig. 4.3   Incidence index and relations of ordered parity

Fig. 4.3 (right) shows an example of a relation for ordered parity.

Remark 4.1. 1. For any ordered parity p and a decreasing Reidemeister move $f: D \to D'$ at the crossings u and v one has

$$p_D(u)p_D(v) = 1,$$

since the incidence indices of the crossings u and v coincide.

2. Oriented parity does not depend on the orientation of the knot. More precise, if a family of maps p is an oriented parity on the diagram category of the knot $\mathcal{K}$, then it is an oriented parity on the diagram category of the knot $-\mathcal{K}$ which is obtained from $\mathcal{K}$ by orientation reversion (we identify naturally here the diagram categories of the knots $\mathcal{K}$ and $-\mathcal{K}$).

3. Parities with coefficients introduced in [29] are all weak parities by definition of parity and equalities $2p(v) = 0$ (see Lemma 4.4 in [29]).

Proposition 4.2. *Let p be an (oriented) parity on a diagram category $\mathfrak{K}$. Then the family of maps $\psi_D: \mathcal{V}(D) \to \mathbb{Z}_2$ defined by the rule*

$$\psi_D(v) = \begin{cases} 1, & p_D(v) \neq 1, \\ 0, & p_D(v) = 1, \end{cases}$$

is a non-trivial weak parity on the diagram category $\mathfrak{K}$. This weak parity is called the induced weak parity *of the parity p.*

Proof. Property $(\Psi 0)$ follows form the first property of oriented parity definition and property $(\Psi 1)$ follows from the second one. Thus ψ is not the trivial weak parity. The third condition of oriented parity definition means that the parities of the crossings involved in a second Reidemeister move are both either equal to the unit of the coefficient group or not equal to it. Thus, $(\Psi 2)$ holds. Analogously, if two crossings involved in a third Reidemeister move are even with respect to p, then the third crossing is even too. So property $(\Psi 3)$ is valid. □

4.2.2 Operations on weak parities

One can define a natural partial ordering on weak parities.

Definition 4.4. Let ψ, ψ' be weak parities on a diagram category $\mathfrak{K}$. We say that ψ *is not greater than* ψ' ($\psi \leqslant \psi'$) if for any diagram $D \in \mathrm{ob}(\mathfrak{K})$ and any vertex $v \in \mathcal{V}(D)$ one has $\psi_D(v) \leqslant \psi'_D(v)$.

Theorem 4.2. *Let $\{\psi_\alpha\}_{\alpha \in A}$ be a family of weak parities on a diagram category $\mathfrak{K}$. Then the family of maps $\psi_D \colon \mathcal{V}(D) \to \mathbb{Z}_2$, $D \in \mathrm{ob}(\mathfrak{K})$, where $\psi_D(v) = \max_{\alpha \in A}(\psi_\alpha)_D(v)$ for every crossing $v \in \mathcal{V}(D)$, is a weak parity. If all the weak parities ψ_α, $\alpha \in A$, are non-trivial, then the parity ψ is non-trivial.*

Proof. We need to check properties $(\Psi 0) - (\Psi 3)$ for ψ.

$(\Psi 0)$. Let $f \colon D \to D'$ be a Reidemeister move and let $v \in \mathcal{V}(D_1)$ be not involved in this move. Then for any $\alpha \in A$ we have $(\psi_\alpha)_D(v) = (\psi_\alpha)_{D'}(f_*(v))$. Then

$$\psi_D(v) = \max_{\alpha \in A}(\psi_\alpha)_D(v) = \max_{\alpha \in A}(\psi_\alpha)_{D'}(f_*(v)) = \psi_{D'}(f_*(v)).$$

$(\Psi 2)$. Let $f \colon D \to D'$ be a decreasing second Reidemeister move at crossings $u, v \in \mathcal{V}(D)$. Then $(\psi_\alpha)_D(u) = (\psi_\alpha)_D(v)$ for any $\alpha \in A$. Hence,

$$\psi_D(u) = \max_{\alpha \in A}(\psi_\alpha)_D(u) = \max_{\alpha \in A}(\psi_\alpha)_D(v) = \psi_D(v).$$

$(\Psi 3)$. Let $f \colon D \to D'$ be a third Reidemeister move at crossings $u, v, w \in \mathcal{V}(D)$. Then $\psi_D(u) = \psi_{D'}(f_*(u))$, $\psi_D(v) = \psi_{D'}(f_*(v))$, $\psi_D(w) = \psi_{D'}(f_*(w))$. Assume that $\psi_D(u) + \psi_D(v) + \psi_D(w) = 1 \in \mathbb{Z}$. Without loss of generality, we can suppose that $\psi_D(u) = \psi_D(v) = 0$ and $\psi_D(w) = 1$. Then for any $\alpha \in A$ we have $(\psi_\alpha)_D(u) = (\psi_\alpha)_D(v) = 0$. Property $(\Psi 3)$ means for the weak parity ψ_α that $(\psi_\alpha)_D(w) = 0$. Therefore, we have $\psi_D(w) = \max_{\alpha \in A}(\psi_\alpha)_D(w) = 0$, and come to a contradiction.

$(\Psi 1)$. Let all the weak parities ψ_α, $\alpha \in A$, be non-trivial. Consider a diagram D with a crossing v which has appeared after a first Reidemeister move. By Theorem 4.1 $(\psi_\alpha)_D(v) = 0$ for any $\alpha \in A$. Then $\psi_D(v) = 0$ so the weak parity ψ is non-trivial. $\qquad\square$

Definition 4.5. The weak parity ψ in Theorem 4.2 is called the *maximum* of the set of weak parities $\{\psi_\alpha\}_{\alpha \in A}$. It will be denoted as $\max_{\alpha \in A} \psi_\alpha$.

Ordering of weak parities distinguishes maximal and minimal elements among them. Note that the minimal weak parity is the null parity since $\mathbb{O} \leqslant \psi$ for any weak parity ψ. The maximal element among all the weak parities is the trivial weak parity $\mathbb{1}$. We shall consider below only non-trivial weak parities.

Definition 4.6. A non-trivial weak parity ψ on a diagram category $\mathfrak{K}$ is called *maximal* if for any non-trivial weak parity ψ' on $\mathfrak{K}$ the inequality $\psi' \leqslant \psi$ holds. We denote below the maximal non-trivial parity as ψ_{max}.

The following statement can be immediately deduced from Theorem 4.2: one should just take the set of all non-trivial weak parities on $\mathfrak{K}$ as the family $\{\psi_\alpha\}_{\alpha \in A}$.

Corollary 4.1. *On any diagram category $\mathfrak{K}$ there exists a unique maximal non-trivial weak parity.*

Below we give a description of the maximal weak parity ψ_{max} for some diagram categories.

Interpreting weak parities as functorial maps, we can define a multiplication of weak parities.

Definition 4.7. Let ψ, ψ' be weak parities on virtual or flat knots and let Ψ, Ψ' be the corresponding functorial maps. The weak parity $\psi \circ \psi'$ that corresponds to the functorial map $\Psi \circ \Psi'$ is called the *product* of the weak parities ψ, ψ'.

Remark 4.2. It is easy to see that $\psi \circ \psi' \geqslant \psi'$ for any weak parities ψ, ψ'. In particular, this means $\psi \circ \psi_{max} = \psi_{max}$ for any non-trivial weak parity ψ.

4.2.3　*Weak parities for knots in a given surface*

Let $\mathfrak{K}$ be a diagram category of a knot (or a flat knot) $\mathcal{K}$ that lies in a given connected oriented closed surface S.

We define the *homotopical weak parity* ψ^{hom} as follows: for any diagram D and any crossing $v \in \mathcal{V}(D)$ we set $\psi_D^{hom}(v) = 0$ if and only if in the homotopy group $\pi_1(S, v)$ the following equality holds

$$[D_v] = [D]^l \tag{4.1}$$

for some integer l. Here D_v denotes one of halves of the knot at the vertex v (see Fig. 4.4). In other words $\psi_D^{hom}(v) = 0$ if the homotopy class of the half D_v belongs to the subgroup generated by the homotopy class $\mathcal{K}$.

Note that classes $[D_v]$ and $[D]$ in (4.1) are ambiguous: the knot has two halves D_v and D_v' at the vertex v, on the other hand, the homotopy class

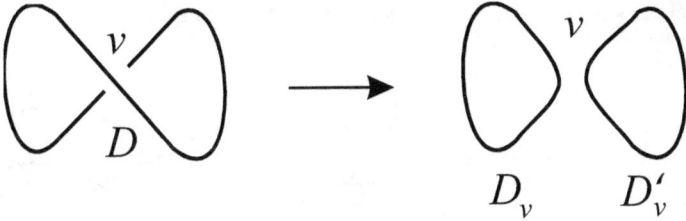

Fig. 4.4   Halves of a knot at a crossing

$[\mathcal{K}]$ equals either $[D_v][D'_v]$ or $[D'_v][D_v]$. Thus, there are four variants of the equation (4.1). Nonetheless, if the equation holds for one variant then it is valid for all other variants (perhaps with different exponent l).

For example, let $[D_v] = ([D_v][D'_v])^l$. Then $[D'_v] = [D_v]^{-1}[D_v][D'_v] = ([D_v][D'_v])^{1-l}$. Hence, elements $[D_v]$ and $[D'_v]$ commutes, so $[D_v][D'_v] = [D'_v][D_v]$.

Proposition 4.3. *The family of maps ψ^{hom} is a weak parity.*

Proof. Let D be a diagram and v be its crossing. The left and right terms of the equation (4.1) do not change by any homotopy of the curve, that is the image of the diagram D, if the homotopy does not involve the crossing v. Therefore, ψ^{hom} possesses the property ($\Psi 0$).

Let us check the property ($\Psi 2$). Let u and v be crossings of the diagram D that take part in a second Reidemeister move. Since the bigon formed by the vertices u and v is contractible, we can identify the homotopy groups $\pi_1(S, u)$ and $\pi_1(S, v)$. At that the halves D_u and D'_u are identified with the halves D_v and D'_v. Hence the equation $[D_u] = [D]^l$ means the same as the equation $[D_v] = [D]^l$. Thus, the parities of the crossings u and v with respect to ψ^{hom} coincide.

Let u, v and w be crossings of the diagram D which take part in a third Reidemeister move. Let α, β and γ be the arcs of the diagram D which connect the crossings u, v and w (see Fig. 4.5). The triangle spanned by the crossings can be contracted to a point $x \in S$. Then we can identify α, β and γ with halves D_u, D_v and D_w respectively, and this correspondence does not depend on how the crossings u, v and w are connected in D. Below α, β and γ will stand for the homotopy classes of the corresponding curves.

These classes are considered as elements of the group $\pi_1(S, x)$.

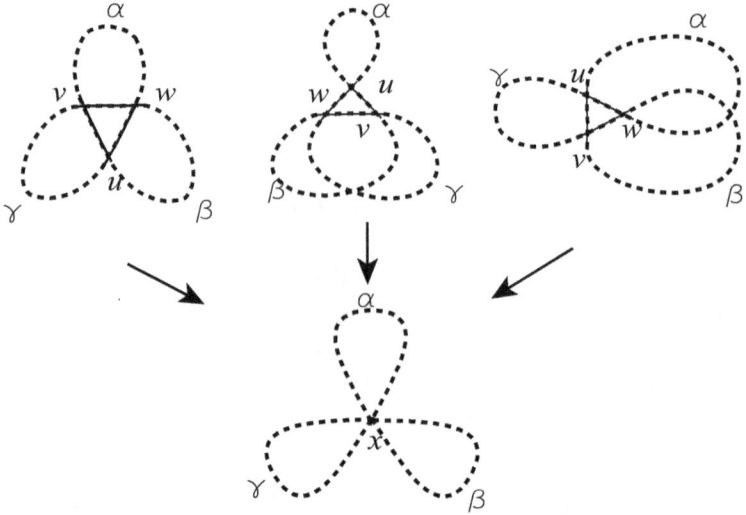

Fig. 4.5   Contracting of the triangle to a point for different connections of its vertices in the diagram D

Assume that there are exactly one odd crossing among u, v and w. Without loss of generality, we can suppose that the odd crossing is w. The halves of the knot at u are equal to α and $\beta\gamma$. Since u is even, we have $\alpha = (\alpha\beta\gamma)^l$ for some $l \in \mathbb{Z}$. Moreover, since the homotopy classes of halves commute, then $\alpha\beta\gamma = \beta\gamma\alpha$. Analogously, for the crossing v we get equalities $\beta = (\beta\gamma\alpha)^m, m \in \mathbb{Z}$, and $\beta\gamma\alpha = \gamma\alpha\beta$. Then

$$\alpha\beta = (\alpha\beta\gamma)^l(\beta\gamma\alpha)^m = (\gamma\alpha\beta)^{l+m},$$

that implies $\gamma = (\gamma\alpha\beta)^{1-l-m}$. Thus, the crossing w must be even. The contradiction shows ψ^{hom} has the property ($\Psi2$), so it is a weak parity. $\square$

Theorem 4.3. *Homotopical parity is the maximal non-trivial weak parity.*

The key part of the proof is the following lemma.

Lemma 4.1. *Let D be a knot diagram in the surface S and $v \in \mathcal{V}(D)$ be its crossing such that $[D_v] = 0 \in \pi_1(S, v)$ for one of the halves of the knot at v. Then for any non-trivial weak parity ψ we have $\psi_D(v) = 0$.*

Proof. If the knot $\mathcal{K}$ is flat, i.e., in the diagram category $\mathfrak{K}$ one does not distinguish undercrossings and overcrossings, then a homotopy of the half D_v to a point can be presented as a sequence of Reidemeister moves. Hence, the half D_v can be transformed by Reidemeister moves to a small loop in a neighborhood of the crossing v (see Fig. 4.6). Then $\psi(v)$ must be equal to 0 due to property (Ψ1).

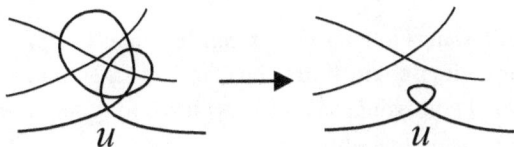

Fig. 4.6   Proof of lemma for a flat knot in the surface

The proof of lemma for virtual (non-flat) knots requires several auxiliary statements.

Lemma 4.2. *For any non-trivial weak parity*

- *the parities of the vertices of an alternating bigon coincide;*
- *the number of odd vertices in an alternating triangle equals 0, 2 or 3.*

Fig. 4.7   Alternating bigon and triangle

Proof. Let crossings u and v form an alternating bigon. With a first Reidemeister move add a crossing w in an arc which connects u and v, so that one could apply the third Reidemeister move to the crossings u, v and w (see Fig. 4.8). The crossing w is even by property (Ψ1), so the parities of the crossings u and v coincide by property (Ψ3).

Let crossings u, v and w confine an alternating triangle. With a second Reidemeister move add crossings t and t' (see Fig. 4.9). Due to (Ψ3), the

Fig. 4.8   Proof for an alternating bigon

number of odd crossings among v, w t can be equal to 0, 2 and 3. But the parities of the crossings t and t' are the same by ($\Psi2$), and the coincidence of the parities of the crossings t' and u, which bound an alternating bigon, follows from the reasonings above.

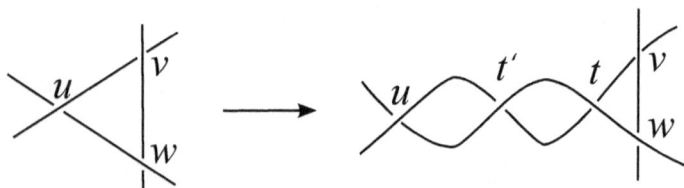

Fig. 4.9   Proof for an alternating triangle

Thus lemma 4.2 is proved.   □

Lemma 4.3. *For any non-trivial weak parity*

- *the parities of the vertices u, v of a crossed bigon (see Fig. 4.10, left) coincide;*
- *the vertex u in a crossed loop (see Fig. 4.10, right) is even*

regardless the choice of over/undercrossings in the vertices.

Proof. Let us consider a crossed bigon with vertices u, v. If the crossing arc can be removed from the bigon by a third Reidemeister move, the coincidence of the parities of u and v follows from property ($\Psi0$) and either property ($\Psi2$) or Lemma 4.2.

Assume now that the crossing arc cannot be removed (see Fig. 4.11). Add crossings w and w' with a second Reidemeister move. The reasonings above show that the parities of the crossings u and w coincide. On the

Fig. 4.10   Crossed bigon and loop

other hand, the parities of w and w' coincide due to property ($\Psi 2$) and the parities of w' and v are the same due to Lemma 4.2. Thus, the parities of the crossings u and v are equal.

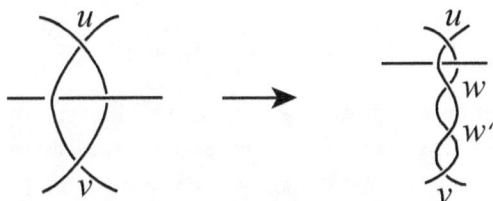

Fig. 4.11   Proof for a crossed bigon

Let us consider a crossed loop with a vertex u, see Fig. 4.10 (right). With a first Reidemeister move we can add a crossing v in the top part of the loop. The crossing v is even by property ($\Psi 1$). The crossings u and v form a crossed bigon, hence, their parities coincide. Thus, the crossing u is even. Lemma 4.3 is proved. □

The following lemma is an analogue of Lemma 4.7 from [29].

Lemma 4.4. *Let a diagram D bound in the surface a polygon whose crossings are $v_0, v_1, \ldots, v_n$. Let the crossings $v_1, \ldots, v_n$ be even with respect to a weak parity ψ. Then the crossing v_0 is also even with respect to ψ.*

Proof.   We prove the lemma by induction. The cases $n = 0, 1, 2$ follow from properties ($\Psi 1$), ($\Psi 2$) and ($\Psi 3$) (as well as Lemma 4.2), respectively.

Assume the statement of the lemma is valid when $n \leqslant k$. Take $n = k+1$. Consider a $(k + 2)$-gon $v_0, v_1, \ldots, v_{k+1}$ such that the crossings $v_1, \ldots, v_{k+1}$

are even. Apply an increasing second Reidemeister move to the edges v_0v_1 and v_kv_{k+1}. Let w and w' be the crossings appeared after the move. The $(k+2)$-gon splits into triangle v_0wv_{k+1}, bigon ww' and $(k+1)$-gon $w'v_1 \ldots v_k$ (see Fig. 4.12).

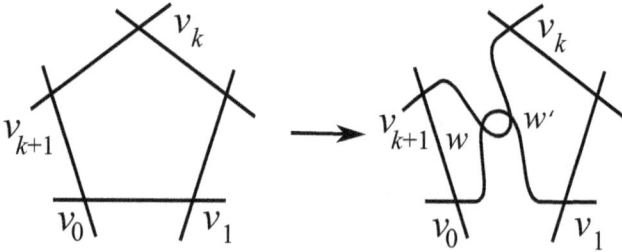

Fig. 4.12   Proof of lemma 4.4

By the induction assumption the crossing w' is even. By property ($\Psi2$) the crossing w is also even. Since the crossing v_{k+1} is even, the remaining vertex v_0 in the triangle v_0wv_{k+1} must be even by property ($\Psi3$) or Lemma 4.2.

Lemma 4.4 is proved. $\square$

Let us finish the proof of Lemma 4.1.

Consider a crossing u of the diagram D with a contractible half D_u (see Fig. 4.13).

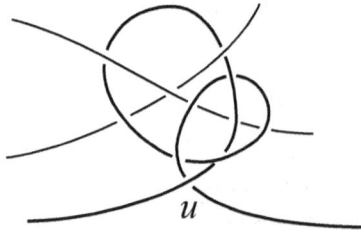

Fig. 4.13   Homotopically trivial half

The contraction process can be considered as a homotopy of the knot in the thickening $S \times I$ of the surface S. The homotopy is represented as a sequence of knot isotopies in $S \times I$ and self-intersections which look after projection like a change of under- and overcrossing in some crossing (see Fig. 4.14).

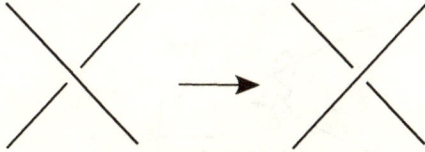

Fig. 4.14   Crossing switch

We modify the homotopy at any crossing switch by adding a small loop along a segment that connects the two arcs of the knot near the crossing to avoid self-intersections (see Fig. 4.15).

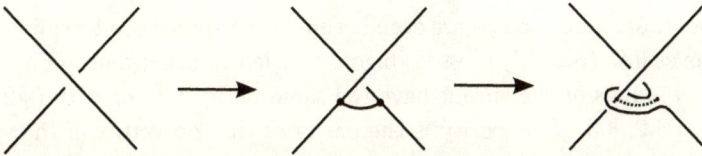

Fig. 4.15   Adding a loop instead of crossing change

As a result we get an isotopy of the knot whose projection transforms the half D_u to a small loop with "sprouts" (see Fig. 4.16).

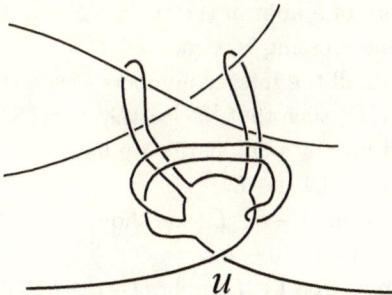

Fig. 4.16   Contracted half with sprouts

Then we apply second Reidemeister moves to get a small polygon which contains the crossing u and the remaining vertices of it correspond to "sprouts" (see Fig. 4.17).

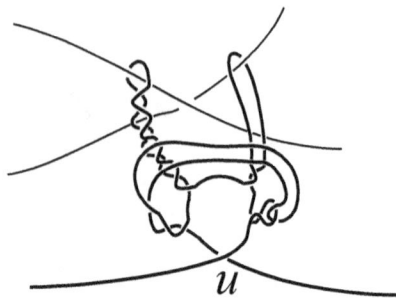

Fig. 4.17 Adding crossings at the bottom of sprouts

Note that those vertices are even because each sprout can be split by second moves into (possibly crossed) bigons (see left upper sprout in Fig. 4.17), so the crossings of the sprout have the same parity by property ($\Psi 2$) and Lemmas 4.2, 4.3. This parity is the parity of the top vertex of the sprout which is the vertex of a crossed loop and therefore is even. Thus, u is a vertex in a polygon whose other vertices are even. Then by Lemma 4.4 the crossing u is even.

Lemma 4.1 is proved. $\qquad\qquad\square$

Proof of Theorem 4.2. It is enough to show that for any weak parity ψ on diagram category of a knot in the surface S the relation (4.1) implies that the corresponding crossing is even.

Let L be the set of all the integer numbers l such that for any diagram D and crossing $v \in \mathcal{V}(D)$ such that $[D_v] = [D]^l \in \pi_1(S, v)$ for some half D_v of the knot at v, and for any weak parity we have $\psi_D(v) = 0$.

According to Lemma 4.1 $0 \in L$.

Obviously, $l \in L$ means $1 - l \in L$ (one should consider the other half at the crossing). Hence, $1 = 1 - 0 \in L$.

We demonstrate now that $l \in L$ implies $l - 1 \in L$. Let $l \in L$ and we have $[D_v] = [D]^{l-1}$ for a crossing v. Add a loop with homotopy type $[D_u] = [D]$ near v, see Fig. 4.18 (we identify here fundamental groups with basepoints in a contractible neighborhood of v). The parity of v does not change by property ($\Psi 0$). Apply a second Reidemeister move and add a crossing w. The homotopy type of some half of the knot at the crossing w is

$$[D_w] = [D_v][D_u] = [D]^{l-1}[D] = [D]^l.$$

Since $l \in L$, the crossing w is even. For $1 \in L$, the crossing u is even too. But the crossings u, v and w for a triangle. Therefore, by property ($\Psi 3$) or Lemma 4.2 the crossing v is even.

Fig. 4.18   Deduction $l \mapsto l - 1$

Thus, all integers which are not grater than 1 belong to L. Then $\mathbb{N} \in L$, since for any $n \in \mathbb{N}$ we have $n = 1 - (1 - n)$ and $1 - n < 1$. Hence, $L = \mathbb{Z}$.

Thus, any crossing, which is even for the homotopical parity, is even for any non-trivial weak parity. Therefore, the homotopical parity ψ^{hom} is maximal. $\square$

Corollary 4.2. *Any weak parity on classical knots is null or trivial.*

Proof. Classical knots are knots in the sphere. The fundamental group of the sphere is trivial, so all crossings are even for the homotopical parity, i.e., $\psi^{hom} = \mathbb{O}$. Since the homotopical parity is maximal, any non-trivial weak parity coincides with the null parity. $\square$

Thus, theory of weak parities becomes trivial for classical knots. An analogous statement about parities was proved in Corollary 4.2 from [29].

Corollary 4.2 can be reformulated in terms of functorial maps as follows.

Corollary 4.3. *The restriction of any non-trivial functorial map to classical knots is the identity. Particularly, any non-trivial map of virtual knots into classical knot diagrams is a projection on classical knots.*

We describe below the weak parities on knots in a given two-dimensional surface. We impose an additional restriction on diagrams — we require that diagram goes through a fixed point in the surface.

Let $\mathcal{K}$ be a knot in the thickening of a connected oriented closed surface S and z be a point in the surface S. Let us consider the diagram category $\mathfrak{K}_z$ whose objects are diagrams of $\mathcal{K}$ which contain z so that z is not a crossing. The morphisms of the category are isotopies and Reidemeister moves which keep the point z fixed.

Theorem 4.4. *There is a natural bijection between weak parities on the diagram category $\mathfrak{K}_z$ and subgroups of the fundamental group $H \subset \pi_1(S, z)$ such that $[\mathcal{K}] \in H$.*

Given a subgroup H, the correspondent weak parity ψ^H is defined as follows: a crossing v of a diagram D is even (i.e., $\psi_D^H(v) = 0$) if and only if $[\hat{D}_v] \in H$ where $\hat{D}_v$ is the based half of the diagram at the crossing v (see Fig. 4.19).

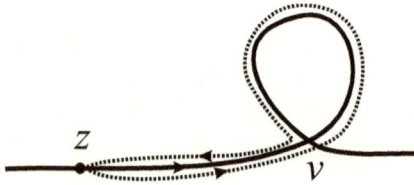

Fig. 4.19   Based half of the diagram at v

Remark 4.3. The definition of the based half depends on the orientation of the knot but the parity of crossings is the same for both orientations as we show below in the proof of the theorem.

Proof. Let H be a subgroup in $\pi_1(S, z)$ that contains the element $[\mathcal{K}]$. Let us show that the family of maps $\psi^H = \{\psi_D^H\}$, $\psi_D^H : \mathcal{V}(D) \to \mathbb{Z}_2$, $D \in \mathrm{ob}(\mathfrak{K}_z)$, defined as follows

$$\psi_D^H(v) = \begin{cases} 0, & [\hat{D}_v] \in H, \\ 1, & \text{otherwise}, \end{cases}$$

is a weak parity.

Let us show that the parity of a crossing does not depend on the choice of the half at the crossing and on the orientation of the knot.

Consider a crossing v and denote the segment of the knot which connects (along the orientation) the points z and v, by γ_1; denote the (unbased) half of the knot at v, which does not contain z, as α; denote the segment of the knot which connects (along the orientation) v and z, by γ_2 (see Fig. 4.20).

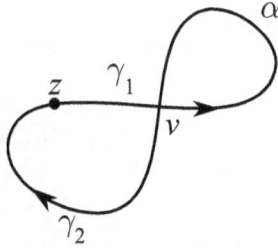

Fig. 4.20   A crossing in a diagram based at z

Then the homotopy class of one based half at the crossing v is $[\hat{D}_v] = [\gamma_1\alpha\gamma_1^{-1}]$ and the class of the other half is $[\hat{D}'_v] = [\gamma_1\alpha\gamma_2\gamma_1\alpha^{-1}\gamma_1^{-1}] = [D][\hat{D}_v^{-1}]$, where $[D] = [\gamma_1\alpha\gamma_2]$ is the homotopy class of the diagram. Since $[D] \in H$, the halves $[\hat{D}_v]$ and $[\hat{D}'_v]$ either belong or do not belong to the subgroup H simultaneously.

For the opposite orientation of the knot the based half of the diagram equals $(-\hat{D})_v = \gamma_2^{-1}\alpha^{-1}\gamma_2$. Then we have $[(-\hat{D})_v] = [D]^{-1}[\hat{D}_v]^{-1}[D]$. Since $[D] \in H$, the condition $[\hat{D}_v] \in H$ is equivalent to $[(-\hat{D})_v] \in H$.

Let us check the properties of parity.

The property ($\Psi 0$) follows from the fact that the homotopy class of the half at a crossing does not change under isotopy and Reidemeister moves that keep the crossing. Nonetheless one should check the situation when the crossing v goes through the point z (see Fig. 4.21).

Let α and β be the (unbased) halves of the diagram at the crossing v. The short arc, which connects v and z, can be contracted. At that the halves before the crossing gets over z and the halves after that can be identified and considered as loops with the base point z. The reduced half $\hat{D}_v$ in the original diagram is equal to α, after the crossing has got over z the corresponding based half is $\hat{D}'_v = \alpha\beta\alpha\beta^{-1}\alpha^{-1}$ (see Fig. 4.21). Then $[\hat{D}'_v] = [D][\hat{D}_v][D]^{-1}$. Since $[D] \in H$, the parity of the crossing remains unchanged.

Let us check the property ($\Psi 2$). Let crossings u and v take part in a second Reidemeister move (see Fig. 4.22). Then $\hat{D}_u = \gamma_1\delta_1\alpha\delta_2\gamma_1^{-1}$ and

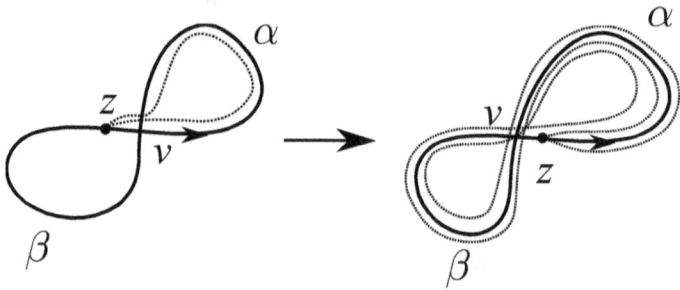

Fig. 4.21   A crossing goes through the base point z

$\hat{D}_v = \gamma_1 \delta_1 \alpha \delta_1^{-1} \gamma_1^{-1}$. Since the small arcs δ_2 and δ_1^{-1} are homotopic as paths with fixed endpoints, we have $[\hat{D}_u] = [\hat{D}_v]$. So the crossings u and v have the same parity with respect to ψ^H.

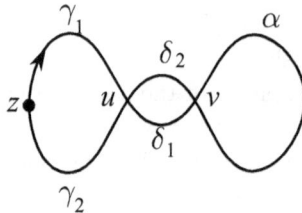

Fig. 4.22   Proof of (Ψ2)

Let crossings u, v and w take part in a third move (see Fig. 4.23). Their based halves are $\hat{D}_u = \gamma_1 \alpha \gamma_1^{-1}$, $\hat{D}_v = \gamma_1 \alpha \beta \alpha^{-1} \gamma_1^{-1}$ and $\hat{D}_w = \gamma_1 \alpha \beta \gamma_2 \gamma_1 \beta^{-1} \alpha^{-1} \gamma_1^{-1}$. Hence, $[\hat{D}_w][\hat{D}_v][\hat{D}_u] = [\gamma_1 \alpha \beta \gamma_2] = [D] \in H$. The equality means that the number of the halves, whose homotopy classes belong to the subgroup H, can be 0, 1 or 3. In other words, the number of odd crossings among u, v and w cannot be equal to 1, i.e., the (Ψ3) holds. Verification of the property (Ψ3) for other configurations of crossings can be done analogously.

Thus, we have shown that ψ^H is a weak parity. Let us demonstrate now that any weak parity on the diagram category $\mathfrak{K}_z$ is equal to ψ^H for some subgroup H.

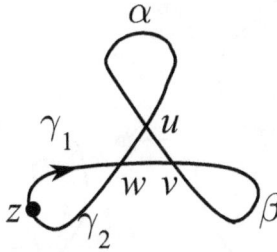

Fig. 4.23   Proof of ($\Psi 3$)

Let ψ be a weak parity on $\mathfrak{K}_z$.

Assume that a diagram D contains crossings u and v such that $[\hat{D}_u] = [\hat{D}_v] \in \pi_1(S, z)$. Let us show that $\psi_D(u) = \psi_D(v)$. To this end we pull the crossings u and v to the base point (see Fig. 4.24), so that the crossings lie in a contractible neighbourhood of the point z.

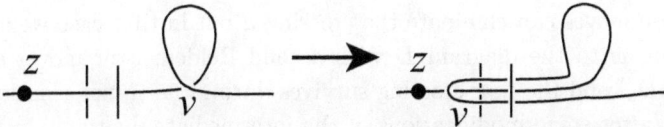

Fig. 4.24   Drawing a crossing to the base point

Note that drawing of a crossing does not change the homotopy type of its half and its parity.

Denote the half of the crossing u, which does not contain z, by α, the long arc that connects the crossings u and v by β, and the long arc that connects v and z by γ (see Fig. 4.25).

We shall identify points inside the contractible neighbourhood, ignore arcs that lie entirely in it and consider the curves α, β, γ as three loops based at z. Then the based half at the crossing u is equal to $\hat{D}_u = \alpha$ and the half at v is $\hat{D}_v = \alpha\beta$. Since we assume $[\hat{D}_u] = [\hat{D}_v]$, then $[\beta] = 1 \in \pi_1(S, z)$. Apply a second Reidemeister move that adds crossings w and w'. One of the (unbased) halves at w is β, so it is contractible. By Lemma 4.1 we have $\psi_D(w) = 0$. Since the crossings u, v and w form a triangle, the property

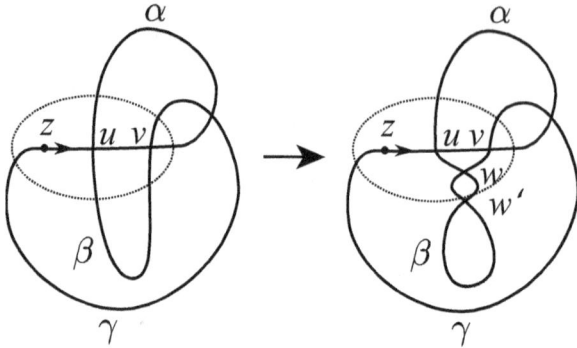

Fig. 4.25 Diagram with the crossings u and v drawn to the base point. The dotted line marks a contractible neighbourhood of the point z

($\Psi3$) and Lemma 4.2 imply that the crossings u and v have the same parity.

Now, let u be a crossing of D and a v be a crossing of D', $D, D' \in \mathrm{ob}(\mathfrak{K}_z)$, such that $[\hat{D}_u] = [\hat{D}'_v] \in \pi_1(S, z)$. Since the diagrams D and D' correspond to the same knot $\mathcal{K}$, there is a sequence of moves which transforms D into D'. These moves can eliminate the crossing u but in that case we can add a crossing w to the diagram D with second Reidemeister moves so that $[\hat{D}_w] = [\hat{D}_u]$ and the new crossing survives during the transformation of D into D' (after some modifications of the intermediate diagrams by adding the crossing w to them with second and perhaps third Reidemeister moves). The homotopy type of the half at w and the parity of the crossing w do not change under moves, so $\psi_D(u) = \psi_D(w) = \psi_{D'}(w) = \psi_{D'}(v)$ where the first and the last equality follow from the reasonings above.

Let $H \subset \pi_1(S, z)$ be the set of homotopy classes of based halves at even crossings (with respect to the weak parity ψ):

$$H = \{[\hat{D}_v] \mid D \in \mathrm{ob}(\mathfrak{K}_z), v \in \mathcal{V}(D) \text{ such that } \psi_D(v) = 0\}.$$

Then the reasoning above means that the weak parity ψ can be defined as follows

$$\psi_D(v) = \begin{cases} 0, & [\hat{D}_v] \in H, \\ 1, & \text{otherwise.} \end{cases}$$

The property ($\Psi1$) implies that $[\mathcal{K}] \in H$. We need to show that H is a subgroup in $\pi_1(S, z)$.

Let $h_1, h_2 \in H$. For any diagram D we can construct with second Reidemeister moves crossings u and v such that $[\hat{D}_u] = h_1$, $[\hat{D}_v] = h_2$. Then $\psi_D(u) = \psi_D(v) = 0$. Draw the crossings u and v to the base point z and add with a second move crossings w and w' (see Fig. 4.25). Then we have $\hat{D}_u = \alpha$, $\hat{D}_v = \alpha\beta$, $\hat{D}_w = \alpha\beta\alpha^{-1}$ (here we ignore the arc inside a contractible neighbourhood of z again). Hence, $[\hat{D}_w] = [\hat{D}_v][\hat{D}_u]^{-1} = h_2 h_1^{-1}$. By the property ($\Psi 3$) (and Lemma 4.2) the crossing w is even with respect to ψ, so $h_2 h_1^{-1} \in H$. Since we can take any elements $h_1, h_2 \in H$, the set H is a subgroup of $\pi_1(S, z)$. Thus, $\psi = \psi^H$. $\qquad \square$

Remark 4.4. A weak parity, defined on the diagrams of a knot $\mathcal{K}$ which contain a fixed base point z, can be extended to arbitrary diagrams of the knot as follows. Fix a diagram $D_0 \in \mathfrak{K}_z$. Let D be another diagram of $\mathcal{K}$. Then there is an isotopy in the thickening of the surface whose projection on the surface transforms the diagram D_0 into D. The projection maps the point z, considered as a point of the diagram D_0, to a curve γ that begins at z and ends at some point $w \in D$.

Let the weak parity ψ on $\mathfrak{K}_z$ be determined by a subgroup $H \subset \pi_1(S, z)$. The path γ induces an isomorphism of the fundamental groups $\gamma_* \colon \pi_1(S, z) \to \pi_1(S, w)$. The value of the weak parity ψ on a crossing v of the diagram D can be defined as follows: $\psi_D(v) = 0$ if and only if $[\hat{D}_v] \in \gamma_*(H) \subset \pi_1(S, w)$, where $\hat{D}_v$ is the based half of the knot at the crossing v with the base point w.

Let $\mathcal{K}$ be a knot in the surface S and D be a diagram $\mathcal{K}$ which contains z. Let H be a subgroup of $\pi_1(S, z)$ that contains the homotopy class $[\mathcal{K}]$. The subgroup H determines a covering $p_H \colon \widetilde{S} \to S$ of the surface S. The lifting $\widetilde{D}$ of the diagram D which contains the base point $\widetilde{z}$ of the surface $\widetilde{S}$ is closed since $[\mathcal{K}] \in H$. Then $\widetilde{D}$ is a knot diagram in the surface $\widetilde{S}$. We can describe the weak parity ψ^H and the corresponding functorial map Ψ^H in terms of the covering p_H.

Proposition 4.4. 1) *A crossing v of the diagram D is even with respect to ψ^H if and only if the lifting $\widetilde{D}_v \subset \widetilde{D}$ of the half D_v is a closed curve.*

2) *The lifting $\widetilde{D}$ of the diagram is the image of the diagram D under the functorial map Ψ^H.*

Remark 4.5. In the second statement a neighbourhood of the diagram $\widetilde{D}$

in $\widetilde{S}$ is considered as a surface $S'(\widetilde{D})$ where the virtual knot presented by the diagram $\widetilde{D}$ lies (see Fig. 4.26).

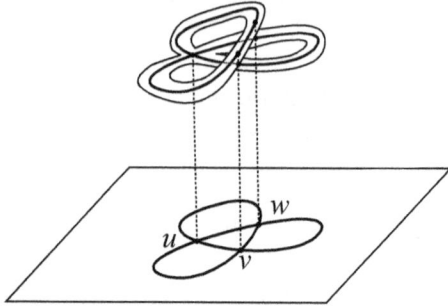

Fig. 4.26   Diagram and it lifting

Proof.   For any crossing we have

$$\psi_D^H(v) = 0 \quad \Longleftrightarrow \quad [\hat{D}_v] \in H \quad \Longleftrightarrow \quad \widetilde{D}_v \text{ is closed.}$$

The second statement follows from the first one and the description of virtual diagrams as diagrams in surfaces. $\square$

For the homotopical weak parity the description of the functorial map as a covering lifting leads to the following statement.

Theorem 4.5. *Let $\mathcal{K}$ be a knot in a connected closed oriented surface S and K be its diagram. Then $\Psi^{hom}(K)$ (where Ψ^{hom} is the functorial map that corresponds to the homotopical weak parity) is a classical knot diagram.*

Proof.   If S is a sphere then K is a classical diagram and the functorial map Ψ^{hom} is the identity.

Let the genus of S is greater that zero. The functorial map $\Psi^{hom}(K)$ can be identified with the lifting of the diagram K in the covering surface $\widetilde{S}$. For the homotopical parity the fundamental group $\pi_1(\widetilde{S})$ is isomorphic to the cyclical subgroup in $\pi_1(\widetilde{S})$ generated by the element $[\mathcal{K}]$. Since the universal covering of S is a plane, the surface $\widetilde{S}$ is a factorization of the plane by a homeomorphism that acts on the plane freely and can be considered to be an isometry of a metric of constant curvature. Therefore,

$\widetilde{S}$ is a plane or cylinder. In any case $\widetilde{S}$ and any neighbourhood of $\Psi^{hom}(K)$ can be embedded into $\mathbb{R}^2$. Thus, $\Psi^{hom}(K)$ is a classical diagram. $\qquad\square$

Remark 4.6. Theorem 4.5 holds also for flat knots in a given surface. It means the functorial map that corresponds to the homotopical weak parity, maps any flat knot to the unknot, since any classical flat knot is trivial.

Corollary 4.4. *We can define a projection* Π *of virtual knots onto classical ones: for any virtual knot* $\mathcal{K}$ *define*

$$\Pi(\mathcal{K}) = \Psi^{hom}(\overline{\mathcal{K}}),$$

where $\overline{\mathcal{K}}$ *is the minimal representative of the knot* $\mathcal{K}$.

Remark 4.7. The projection above is defined on the level of knots but not on the level of diagrams. In other words, the projection is not a functorial map of virtual knots to classical ones. For example, one can take the conventional trefoil diagram with three crossing and attach four handles which connect each bounded region the diagram splits the plane into with the unbounded region. The new diagram will lie in a surface of genus 4, and all the crossings of it will be odd for the homotopical weak parity. Thus, the homotopical functorial map transforms the diagram to a pure virtual diagram of the unknot.

In [54] the projection of the Corollary 4.4 was extended to arbitrary virtual diagrams and a family of maps from virtual diagrams to classical ones, each of which is well defined on knots but is not functorial, was constructed. Let us describe briefly this construction.

Given two virtual diagrams D and D', the diagram D' is called *smaller* than D if it can be obtained form D by replacing some classical crossings of the diagram with virtual ones. For this, we take the notation $D' < D$.

Proposition 4.5. *Let* D *be a virtual knot diagram, whose genus is not minimal in the class of the knot. Then there exists a diagram* $D' < D$ *in the same knot class.*

Proof. Let D be a virtual knot diagram on a surface S_g of genus g. Assume that this genus is not minimal for the knot class of D. Then from Kuperberg's theorem it follows that there is a diagram $\tilde{D}$ on S_g representing the same knot as D and a curve γ on S_g such that $\tilde{D}$ does not intersect γ. Indeed, if there were no such diagram $\tilde{D}$, the knot in $S_g \times I$ corresponding to the diagram D would admit no destabilisation, and the genus g would be minimal.

The curve γ gives rise to a (co)homological parity for knots in S_g homotopic to D: a crossing is *even* if the number of intersections of any of the corresponding halves with γ is even, and *odd*, otherwise.

Since D has underlying diagram genus g, there exists at least one odd crossing of the diagram D. Let D' be the result of γ-parity projection applied to D. We have $D' < D$.

By construction, all crossings of $\tilde{D}$ are even.

Let us construct a chain of Reidemeister moves from D to $\tilde{D}$ and apply the γ-parity projection to it.

We shall get a chain of Reidemeister moves connecting D' to $\tilde{D}$. So, D' is of the same type as $\tilde{D}$ and D. The claim follows.   □

Theorem 4.6. *For every virtual diagram D there exists a classical diagram $\bar{D}$ such that*

1) *$\bar{D} < D$;*
2) *$\bar{D} = D$ if and only if D is classical;*
3) *if D_1 and D_2 are equivalent virtual diagrams, then so do $\bar{D}_1$ and $\bar{D}_2$;*
4) *the map restricted to non-classical knots is a surjection onto the set of all classical knots.*

Proof. We shall construct the projection map in two steps.

Let D be a virtual knot diagram. If D is of a minimal genus, then we take $\bar{D}$ to be just $\Pi(D)$ as in Corollary 4.4. Otherwise take a diagram D' instead of D as in Proposition 4.5. It is of the same knot type as D. If the genus of the resulting diagram is still not minimal, we proceed by iterating the operation D', until we get to a diagram D'' of minimal genus which represents the class of D and $D'' < D$. Now, set $\bar{D} = \Pi(D'')$.

One can easily see that if we insert a small classical knot L inside an edge of a diagram of D, then $\overline{D\#L} = \bar{D}\#\bar{L}$. So, the last statement of the theorem holds as well.   □

From Theorem 4.6 we have the following two corollaries.

Corollary 4.5. *Let $\mathcal{K}$ be an isotopy class of a classical knot. Then the minimal number of classical crossings for virtual diagrams of $\mathcal{K}$ is realised on classical diagrams (and those obtained from them by the detour move). For every non-classical diagram realising a knot from $\mathcal{K}$, the number of classical crossings is strictly greater than the minimal number of classical crossings.*

Moreover, minimal classical crossing number of a non-classical virtual knot is realised only on minimal genus diagrams.

Indeed, the projection map from the theorem above decreases the number of classical crossings, and preserves the knot type.

The observation that the following corollary is a consequence from Theorem 4.6 is due to V. V. Chernov (Tchernov). Let us recall that a *bridge* of a virtual diagram is a branch of the diagram from an undercrossing to the next undercrossing that contains only overcrossings and virtual crossings. The *bridge number $br(K)$* of a virtual knot K is the minimal number of bridges which can have its diagram.

Corollary 4.6. *Let $\mathcal{K}$ be a classical knot. Then the bridge number for $\mathcal{K}$ can be realised on classical diagrams of $\mathcal{K}$ only.*

Moreover, minimal bridge number of a non-classical virtual knot is realised on minimal genus diagrams (here we do not claim that it can not be realised on non-classical diagrams).

Proof. Indeed, it suffices to see that if $D' < D$ then the number of bridges cannot be increased; it can only decrease because two bridges can be joined to form one bridge. $\square$

4.3 Brackets

4.3.1 *The parity bracket*

The first example of the parity bracket firstly appeared in [47]. That bracket was constructed for the Gaussian parity and played a significant role in, for instance, proving minimality theorems, reducing problems about diagrams to questions about graphs etc. Also the bracket was generalised for the case of graph-links, see [26], and allowed the authors to prove the existence of non-realisable graph-links, for more details see [26].

In this section we consider the parity bracket for any parity valued in $\mathbb{Z}_2$. This bracket is a generalisation of the bracket from [47]. Let $\mathfrak{G}$ be the set of all equivalence classes of framed graphs with one unicursal component modulo second Reidemeister moves. Consider the linear space $\mathbb{Z}_2\mathfrak{G}$.

Let $\mathcal{K}$ be a virtual (resp., flat, free) knot, p be a parity on diagrams of $\mathcal{K}$ with coefficients from the group $\mathbb{Z}_2$, and K be a diagram of $\mathcal{K}$ with $\mathcal{V}(K) = \{v_1, \ldots, v_n\}$. For each element $s \in \{0,1\}^n$ we define K_s to be equal to the sum of all graphs obtained from K by a smoothing at each vertex v_i if $s_i = 1$. If $|s| = l$, K_s contains 2^l summands. Define $q_{K,s}(v_i) = p_K(v_i)$ if $s_i = 0$, and $q_{K,s}(v_i) = 1 - p_K(v_i)$ if $s_i = 1$.

Consider the following sum (the *parity bracket*)

$$[K] = \sum_{s \in \{0,1\}^n} \prod_{i=1}^{n} q_{K,s}(v_i) K_s \in \mathbb{Z}_2\mathfrak{G},$$

where only those summands with one unicursal component are taken into account.

Theorem 4.7. *If K and K' represent the same knot then the following equality holds in $\mathbb{Z}_2\mathfrak{G}$:*

$$[K] = [K'].$$

Proof. Let us check the invariance $[K] \in \mathbb{Z}_2\mathfrak{G}$ under the three Reidemeister moves.

1) Let K' differ from K by a first Reidemeister move, and $\mathcal{V}(K') = \{v_1, v_2, \ldots, v_{n+1}\}$, $\mathcal{V}(K) = \{v_1, v_2, \ldots, v_n\}$. We have

$$p_{K'}(v_{n+1}) = 0$$

and

$$[K'] = \left[\bigotimes\right] = \sum_{s \in \{0,1\}^{n+1}} \prod_{i=1}^{n+1} q_{K',s}(v_i) K'_s$$

$$= \sum_{s \in \{0,1\}^n} \prod_{i=1}^{n} q_{K',s}(v_i) \left(p_{K'}(v_{n+1}) \bigotimes + (1 - p_{K'}(v_{n+1})) \left(\bigotimes + \bigotimes \right) \right)$$

$$= \sum_{s \in \{0,1\}^n} \prod_{i=1}^{n} q_{K',s}(v_i) \bigotimes = [K].$$

2) Let K' be obtained from K by a second Reidemeister move adding two vertices, where $\mathcal{V}(K') = \{v_1, v_2, \ldots, v_{n+1}, v_{n+2}\}$ and $\mathcal{V}(K) = \{v_1, v_2, \ldots, v_n\}$. We have

$$p_{K'}(v_{n+1}) + p_{K'}(v_{n+2}) = 0,$$

i.e.,

$$p_{K'}(v_{n+1}) = p_{K'}(v_{n+2}) = 0 \quad \text{or} \quad p_{K'}(v_{n+1}) = p_{K'}(v_{n+2}) = 1,$$

and

$$
\begin{aligned}
[K'] = \left[\vcenter{\hbox{\includegraphics{}}}\right] &= \sum_{s \in \{0,1\}^{n+2}} \prod_{i=1}^{n+2} q_{K',s}(v_i) K'_s \\
&= \sum_{s \in \{0,1\}^n} \prod_{i=1}^{n} q_{K',s}(v_i) \Big(p_{K'}(v_{n+1}) p_{K'}(v_{n+2}) \vcenter{\hbox{\includegraphics{}}} \\
&\quad + p_{K'}(v_{n+1})(1 - p_{K'}(v_{n+2})) \Big(\vcenter{\hbox{\includegraphics{}}} + \vcenter{\hbox{\includegraphics{}}} \Big) \\
&\quad + (1 - p_{K'}(v_{n+1})) p_{K'}(v_{n+2}) \Big(\vcenter{\hbox{\includegraphics{}}} + \vcenter{\hbox{\includegraphics{}}} \Big) \\
&\quad + (1 - p_{K'}(v_{n+1}))(1 - p_{K'}(v_{n+2})) \Big(\vcenter{\hbox{\includegraphics{}}} + \vcenter{\hbox{\includegraphics{}}} + \vcenter{\hbox{\includegraphics{}}} + \vcenter{\hbox{\includegraphics{}}} \Big) \Big) \\
&= \sum_{s \in \{0,1\}^n} \prod_{i=1}^{n} q_{K',s}(v_i) \vcenter{\hbox{\includegraphics{}}} = [K].
\end{aligned}
$$

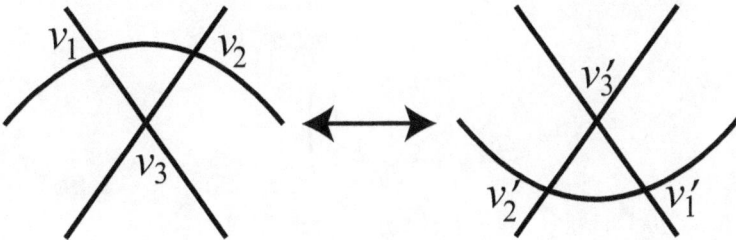

Fig. 4.27 A third Reidemeister move

3) Let K' be obtained from K by a third Reidemeister move applied to vertices v_1, v_2, v_3 in K. Denote by v'_1, v'_2, $v'_3 \in \mathcal{V}(K')$ the vertices corresponding to v_1, v_2, v_3, see Fig. 4.27 (here $\mathcal{V}(K') = \{v_1, v_2, \ldots, v_n\}$ and

$\mathcal{V}(K) = \{v_1', v_2', \ldots, v_n'\})$. We have

$$p_K(v_1) + p_K(v_2) + p_K(v_3) = 0,$$
$$p_{K'}(v_1') + p_{K'}(v_2') + p_{K'}(v_3') = 0,$$

and

$$[K] = \left[\;\vcenter{\hbox{⊗}}\;\right] = \sum_{s \in \{0,1\}^n} \prod_{i=1}^{n} q_{K,s}(v_i) K_s$$

$$= \sum_{s \in \{0,1\}^{n-3}} \prod_{i=4}^{n} q_{K,s}(v_i) \left(\underbrace{p_K(v_1)p_K(v_2)p_K(v_3)}_{=0}\;\vcenter{\hbox{⊗}} \right.$$

$$+ p_K(v_1)p_K(v_2)(1 - p_K(v_3)) \left(\vcenter{\hbox{⊗}} + \vcenter{\hbox{⊗}}\right)$$

$$+ (1 - p_K(v_1))p_K(v_2)p_K(v_3) \left(\vcenter{\hbox{⊗}} + \vcenter{\hbox{⊗}}\right)$$

$$+ p_K(v_1)(1 - p_K(v_2))p_K(v_3) \left(\vcenter{\hbox{⊗}} + \vcenter{\hbox{⊗}}\right)$$

$$+ \underbrace{(1 - p_K(v_1))(1 - p_K(v_2))p_K(v_3)}_{=0} \left(\vcenter{\hbox{⊗}} + \vcenter{\hbox{⊗}} + \vcenter{\hbox{⊗}} + \vcenter{\hbox{⊗}}\right)$$

$$+ \underbrace{(1 - p_K(v_1))p_K(v_2)(1 - p_K(v_3))}_{=0} \left(\vcenter{\hbox{⊗}} + \vcenter{\hbox{⊗}} + \vcenter{\hbox{⊗}} + \vcenter{\hbox{⊗}}\right)$$

$$+ \underbrace{p_K(v_1)(1 - p_K(v_2))(1 - p_K(v_3))}_{=0} \left(\vcenter{\hbox{⊗}} + \vcenter{\hbox{⊗}} + \vcenter{\hbox{⊗}} + \vcenter{\hbox{⊗}}\right)$$

$$+ (1 - p_K(v_1))(1 - p_K(v_2))(1 - p_K(v_3)) \left(\vcenter{\hbox{⊗}} + \vcenter{\hbox{⊗}} + \vcenter{\hbox{⊗}}\right.$$

$$\left.\left. + \vcenter{\hbox{⊗}} + \vcenter{\hbox{⊗}} + \underbrace{\vcenter{\hbox{⊗}}}_{=0} + \underbrace{\vcenter{\hbox{⊗}} + \vcenter{\hbox{⊗}}}_{=0} \right) \right)$$

$$= p_K(v_1)p_K(v_2)(1 - p_K(v_3)) \left(\vcenter{\hbox{⊗}} + \vcenter{\hbox{⊗}}\right)$$

$$+ (1 - p_K(v_1))p_K(v_2)p_K(v_3) \left(\vcenter{\hbox{⊗}} + \vcenter{\hbox{⊗}}\right)$$

$$+ p_K(v_1)(1 - p_K(v_2))p_K(v_3) \left(\vcenter{\hbox{⊗}} + \vcenter{\hbox{⊗}}\right) + (1 - p_K(v_1))$$

$$\cdot (1 - p_K(v_2))(1 - p_K(v_3)) \left(\vcenter{\hbox{⊗}} + \vcenter{\hbox{⊗}} + \vcenter{\hbox{⊗}} + \vcenter{\hbox{⊗}} + \vcenter{\hbox{⊗}}\right),$$

$$[K'] = \left[\;\text{⊗}\;\right] = \sum_{s \in \{0,1\}^n} \prod_{i=1}^{n} q_{K',s}(v_i') K_s'$$

$$= \sum_{s \in \{0,1\}^{n-3}} \prod_{i=4}^{n} q_{K',s}(v_i') \left(\underbrace{p_{K'}(v_1')p_{K'}(v_2')p_{K'}(v_3')}_{=0}\;\text{⊗} \right.$$

$$+ p_{K'}(v_1')p_{K'}(v_2)(1-p_{K'}(v_3')) \left(\text{⊗} + \text{⊗}\right)$$

$$+ (1-p_{K'}(v_1'))p_{K'}(v_2')p_{K'}(v_3') \left(\text{⊗} + \text{⊗}\right)$$

$$+ p_{K'}(v_1')(1-p_{K'}(v_2'))p_{K'}(v_3') \left(\text{⊗} + \text{⊗}\right)$$

$$+ \underbrace{(1-p_{K'}(v_1'))(1-p_{K'}(v_2'))p_{K'}(v_3')}_{=0} \left(\text{⊗} + \text{⊗} + \text{⊗} + \text{⊗}\right)$$

$$+ \underbrace{(1-p_{K'}(v_1'))p_{K'}(v_2')(1-p_{K'}(v_3'))}_{=0} \left(\text{⊗} + \text{⊗} + \text{⊗} + \text{⊗}\right)$$

$$+ \underbrace{p_{K'}(v_1')(1-p_{K'}(v_2'))(1-p_{K'}(v_3'))}_{=0} \left(\text{⊗} + \text{⊗} + \text{⊗} + \text{⊗}\right)$$

$$+ (1-p_{K'}(v_1'))(1-p_{K'}(v_2'))(1-p_{K'}(v_3')) \left(\text{⊗} + \text{⊗} + \text{⊗}\right.$$

$$\left. + \text{⊗} + \text{⊗} + \underbrace{\text{⊗}}_{=0} + \underbrace{\text{⊗}}_{=0} + \text{⊗}) \right)$$

$$= p_{K'}(v_1')p_{K'}(v_2)(1-p_{K'}(v_3')) \left(\text{⊗} + \text{⊗}\right)$$

$$+ (1-p_{K'}(v_1'))p_{K'}(v_2')p_{K'}(v_3') \left(\text{⊗} + \text{⊗}\right)$$

$$+ p_{K'}(v_1')(1-p_{K'}(v_2'))p_{K'}(v_3') \left(\text{⊗} + \text{⊗}\right) + (1-p_{K'}(v_1'))$$

$$\cdot (1-p_{K'}(v_2'))(1-p_{K'}(v_3')) \left(\text{⊗} + \text{⊗} + \text{⊗} + \text{⊗} + \text{⊗}\right).$$

As we consider $\mathbb{Z}_2\mathfrak{G}$ (i.e., up to second Reidemeister moves), we have

$$\text{⊗} = \text{⊗}, \quad \text{⊗} = \text{⊗}, \quad \text{⊗} = \text{⊗}, \quad \text{⊗} = \text{⊗},$$

$$\text{⊗} = \text{⊗}, \quad \text{⊗} = \text{⊗}, \quad \text{⊗} = \text{⊗}, \quad \text{⊗} = \text{⊗},$$

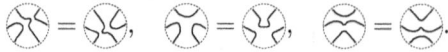

$$\langle\!\langle \rangle\!\rangle = \langle\!\langle \rangle\!\rangle, \quad \langle\!\langle \rangle\!\rangle = \langle\!\langle \rangle\!\rangle, \quad \langle\!\langle \rangle\!\rangle = \langle\!\langle \rangle\!\rangle.$$

Therefore, $[K] = [K']$.   □

4.3.2   *The invariants* $[\cdot]$ *and* $\{\cdot\}$

Let us consider framed 4-graphs with one unicursal component modulo the equivalence relation generated by second Reidemeister moves.

Let us define the linear space $\mathfrak{G}$ as the set of $\mathbb{Z}_2$–linear combinations of such equivalence classes.

The linear space $\widetilde{\mathfrak{G}}$ is the set of $\mathbb{Z}_2$–linear combinations of the following objects. One considers all framed 4-graphs modulo the following equivalence relations:

1) the second Reidemeister move;
2) $L \sqcup \bigcirc = 0$, i.e., a framed 4-graph having more than one component and at least one trivial component, is assumed to be zero.

There is a natural map $g\colon \widetilde{\mathfrak{G}} \to \mathfrak{G}$ which takes to zero all equivalence classes of framed graphs having more than one unicursal component. Obviously, the map g is an epimorphism of groups.

The invariants given below will be explicitly defined for the case of free knots and links with respect to any parity satisfying all parity axioms.

The definitions below can be directly extended to *free tangles*; these objects are encoded by graphs, which, besides vertices of degree 4 (with the usual framing), have free ends of degree 1 (in particular, one can extend the definitions to free braids). We shall not give explicit definitions for the case of free tangles, leaving them for the reader as an exercise.

Since every virtual link generates the free link by "forgetting" the structure at classical crossings (overcrossing/undercrossing structure and the local writhe number), but we remember the structure of opposite edges, then these invariants can be lifted to invariants of virtual knots and links for every parity which is well defined in the corresponding case.

Now we pass to the construction of the invariants $[\cdot]$ and $\{\cdot\}$ of free knots to be valued in $\mathfrak{G}$ and in $\widetilde{\mathfrak{G}}$, respectively.

Let K be a framed 4-graph. Then the invariant $\{\cdot\}$ is given by the formula

$$\{K\} = \sum_{s_{even}} K_{s_{even}} \in \widetilde{\mathfrak{G}},$$

where it follows from the notation that the sum is taken over all even smoothings s_{even} of the framed 4-graph K, which are considered as elements from $\widetilde{\mathfrak{G}}$.

Theorem 4.8. *The bracket $\{\cdot\}$ is an invariant of free links.*

Proof. Let us check the invariance of the bracket $\{\cdot\}$ with respect to the Reidemeister moves. Here we shall use those properties of the parity, satisfied by those crossings of the diagram undergoing the Reidemeister moves.

Let diagrams K and K' differ from each other by one move Ω_1 such that the diagram K' contains one more vertex than the diagram K, and this vertex is denoted by v.

Note that the vertex v is even according to Lemma 2.1. One of the smoothings of the diagram K' at v leads us to a split component in such a way that every even state of the diagram K', where the vertex v is smoothed in the "wrong" way, will lead to a split trivial component. This will yield a trivial element from $\widetilde{\mathfrak{G}}$.

The smoothing of K' at v performed in the "right" way will lead us to the diagram K, which yields a one-to-one correspondence for even smoothings for K and K' with no split circles.

Thus we have proved the invariance of the bracket $\{\cdot\}$ under the first Reidemeister move.

For the second and third Reidemeister moves Ω_2 and Ω_3 we shall show that $\{K\} + \{K'\} \equiv 0 \pmod{\mathbb{Z}_2}$ by means of partitioning of all diagrams from $\{K\} + \{K'\}$ representing non-trivial elements from $\widetilde{\mathfrak{G}}$. Since $\widetilde{\mathfrak{G}}$ is a linear space over $\mathbb{Z}_2$, this will mean that $\{K\} = \{K'\}$.

Let K' be obtained from K by a second Reidemeister move adding two crossings v_1 and v_2. If both crossings are odd, then there is an obvious one-to-one correspondence between the set of even smoothings of the diagram K and the set of even smoothings of the diagram K'. The corresponding smoothings are obtained from each other by applying "the same" Reidemeister move to vertices v_1 and v_2. If both v_1 and v_2 are even, then there exist four smoothings of K' at these vertices:

$$\mathsf{X} \to \mathsf{)(}, \quad \mathsf{X} \to \mathsf{X}, \quad \mathsf{X} \to \mathsf{X}, \quad \mathsf{X} \to \mathsf{X}.$$

Note that the smoothing X has a split circle which will remain split for all subsequent smoothings at even crossings. Thus, such summands will have no impact in $\{K'\}$.

Furthermore, the smoothings X and X are in fact the same framed 4-graph (provided that the smoothings at the remaining vertices agree). Thus, these smoothings cancel in $\{K'\}$.

Even smoothings of type $\mathsf{)(}$ of the diagram K' are naturally in one-to-one correspondence with even smoothings of the diagram K and give rise to framed four-graphs.

Now assume the diagram K is taken to a diagram K' by a third Reidemeister move Ω_3. Among the three crossings of K taking part in the Reidemeister move, either all three ones are even, or one crossing is even, and the other two ones are odd.

If the three crossings of K taking part in the Reidemeister move, are even, then we have seven types of summands in the expansion of $\{K\}$ (and seven types for $\{K'\}$): at each of the three vertices we have two possible smoothings, herewith one of the eight possibilities leads to a split trivial circle for K and one of the eight possibilities for K' leads to a trivial circle (these two types of smoothings do not contribute to $\{K\}$ or to $\{K'\}$). Considering K (in Fig. 4.28 the summands corresponding to K are in the left hand side and the summands corresponding to K' are in the right hand side) we see that three of these seven types lead to coinciding sets of diagrams (these types are denoted by 1), thus in $\widetilde{\mathfrak{G}}$, the two sets are canceled with each other, so, only one set is left. Analogously for the case of K' we have three "similar" types of smoothings (they are denoted by 2 in Fig. 4.28). Thus, both in $\{K\}$ and in $\{K'\}$, five types of summands are left: 1, 2, 3, 4, 5.

In these five cases there is a one-to-one correspondence (see Fig. 4.28), which leads us to the equality $\{K\} = \{K'\}$.

If among the vertices taking part in Ω_3 there is exactly one even vertex in the right hand side and exactly one even vertex in the left hand side (say, $a \to a'$), then we are in the situation shown in Fig. 4.29.

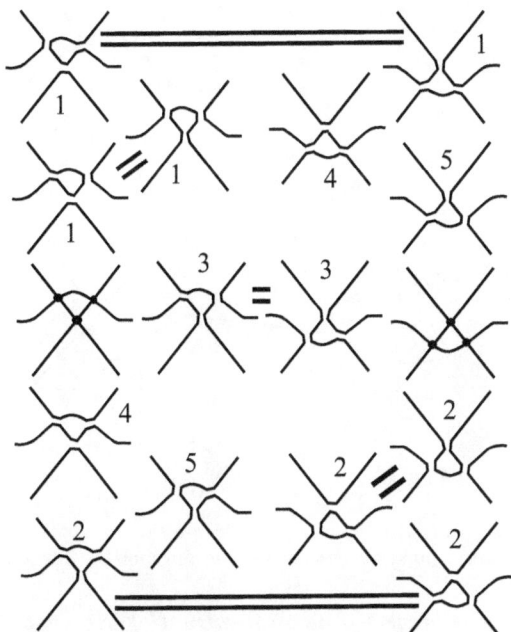

Fig. 4.28 The correspondence between smoothings for Ω_3 with three even vertices

From the above figure we see that those smoothings, where the vertex a (respectively, a') is smoothed *vertically*, give rise to coinciding summands for $\{K\}$ and for $\{K'\}$, and those smoothings, where a and a' are smoothed *horizontally*, are in one-to-one correspondence for diagrams K and K', moreover, the corresponding graphs differ from each other by an application of two second Reidemeister moves. This proves that $\{G\} = \{G'\}$ in $\widetilde{\mathfrak{G}}$. $\qquad\square$

The invariant $[\cdot]$ is given by the formula

$$[K] = \sum_{s_{1,even}} K_{s_{1,even}} \in \mathfrak{G}, \qquad (4.2)$$

where the sum is taken over all even smoothings of the diagram K, which yield one unicursal component.

Obviously, $[K] = g(\{K\})$, which leads to the

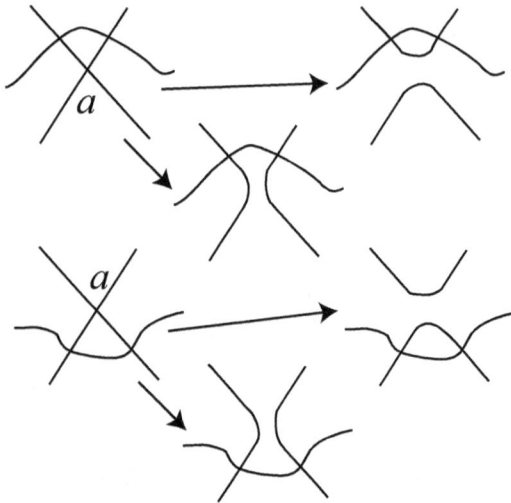

Fig. 4.29　The correspondence between smoothings for Ω_3 with one even vertex

Theorem 4.9. *The bracket* $[\cdot]$ *is an invariant of free knots.*

Sometimes it will be convenient for us to use the bracket $[\cdot]$ just as defined by formula (4.2).

Remark 4.8. Note that the invariants defined above can be constructed for any parity.

The invariants $\{\cdot\}$ and $[\cdot]$ take a certain equivalence class (of diagrams modulo Reidemeister moves) to some linear combinations of equivalence classes (of diagrams modulo the second Reidemeister move and some simple factorization of some class of diagrams). It turns out that the sets $\mathfrak{G}$ and $\widetilde{\mathfrak{G}}$ can be easily described algorithmically: every element of each of these sets has a minimal representative which can be found by means of subsequent simplifications. By a simplification we mean the second Reidemeister move which decreases the number of crossings by 2, and also (in case of elements from $\widetilde{\mathfrak{G}}$) the transformation which takes to zero any diagram with more than one unicursal component having one split trivial component.

More precisely, the above statement can be formulated as a lemma, which we shall formulate after some definition.

We say that a framed 4-graph is *simplifiable*, if it either contains a component without vertices or contains two vertices v_1 and v_2 connected by a couple of edges p, q such that p, q are adjacent both in v_1 and in v_2.

The second case is just the situation when one can apply a second decreasing Reidemeister move to the graph.

Framed 4-graphs not admitting any simplifications, will be called *minimal*. We say that a graph Γ_0 is obtained from a graph Γ by a *subsequent simplification*, if there exists a chain of framed 4-graphs

$$\Gamma = \Gamma_n \to \cdots \to \Gamma_1 \to \Gamma_0,$$

where every subsequent graph is obtained from the previous one by an application of a second simplifying Reidemeister move.

A *minimal representative* of a graph Γ is a minimal graph which can be obtained from Γ by a subsequent simplification.

In the case of $\mathfrak{G}$ the notions of *minimal* and *non-simplifiable* graphs coincide.

Lemma 4.5. *If two framed 4-graphs Γ_0 and Γ_0' both have one unicursal component, and are both obtained from some framed 4-graph Γ by a subsequent simplification, then Γ_0 is isomorphic to Γ_0'.*

The lemma states that every framed 4-graph with one unicursal component has a *unique* minimal representative.

From this lemma one gets the following

Lemma 4.6. *Framed 4-graphs Γ and $\widetilde{\Gamma}$ with one unicursal component each are equivalent in $\mathfrak{G}$ if and only if their minimal representatives coincide.*

Proof. Let us deduce Lemma 4.6 from Lemma 4.5. The claim $\Longleftarrow$ is evident. Assume that $\Gamma = \widetilde{\Gamma}$ is in $\mathfrak{G}$, and here minimal representatives of graphs Γ and $\widetilde{\Gamma}$ are different. Let Δ and Δ' be two framed graphs obtained from each other by a second Reidemeister move in such a way that the graph Δ' has two crossings more than the graph Δ. Then by definition of the minimal representative and by Lemma 4.5 we conclude that the minimal representatives for Δ and Δ' coincide. Considering the chain

$$\Gamma = \Gamma_1 \to \cdots \to \Gamma_k = \widetilde{\Gamma}$$

of the second Reidemeister moves connecting Γ to $\widetilde{\Gamma}$, we see that the minimal representatives of any two adjacent graphs in this chain coincide. Thus the minimal representatives of Γ and $\widetilde{\Gamma}$ coincide as well, which yields the equivalence of these graphs in $\mathfrak{G}$. $\square$

We preface Lemma 4.5 with two simple statements.

Statement 4.1. *Assume that a framed 4-graph Γ with one unicursal component has two non-isomorphic representatives. Then there exists such a simplification $\widetilde{\Gamma}$ of the graph Γ ($\widetilde{\Gamma}$ can coincide with Γ) for which the following holds.*

Among minimal representatives of the graph $\widetilde{\Gamma}$ there exist such non-isomorphic framed 4-graphs Γ_0 and Γ_0', and among elementary simplifications of the framed 4-graph $\widetilde{\Gamma}$ there exist graphs Γ_1 and Γ_1' such that Γ_0 is one of minimal representatives for Γ_1, but not for Γ_1'.

Proof. We choose two non-isomorphic minimal representatives Γ_0 and Γ_0' of the graph Γ and consider the chain of elementary transformations from Γ to Γ_0. In the initial moment the graph Γ has among its minimal representatives the graph Γ_0', and in the end of the chain, among minimal representatives of Γ_0 there is no graph Γ_0' (since the graph Γ_0 cannot be decreased). Consider the chain from Γ to Γ_0. Take the last graph of the chain having Γ_0 as a minimal representative and denote this graph by $\widetilde{\Gamma}$, we get the required statement. $\square$

Statement 4.2. *If framed 4-graphs Γ_1 and Γ_1' are each obtained from a framed 4-graph Γ by one elementary simplification, then either Γ_1 is isomorphic to Γ_1' or there exists a framed 4-graph Γ_2' which can be obtained by one elementary simplification from each of Γ_1, Γ_1'.*

Proof. It suffices to consider the vertices $\{\alpha, \beta\}$ of the graph Γ where the simplification $\Gamma \to \Gamma_1$ takes place, and the vertices $\{\alpha', \beta'\}$ corresponding to the simplification $\Gamma \to \Gamma_1'$. If the set $\{\alpha, \beta\} \cup \{\alpha', \beta'\}$ consists of two or three elements, then it is evident that the graphs Γ_1 and Γ_1' are isomorphic. If this set consists of four elements, then it is easy to see that one can apply an elementary simplification to the graph Γ_1 at $\{\alpha', \beta'\}$ so that the resulting graph is isomorphic to the graph obtained from Γ_1' by an elementary simplification at vertices $\{\alpha, \beta\}$. $\square$

***Proof of Lemma* 4.5.** Consider the framed 4-graph Γ with one unicursal component and assume it has more minimal representatives than one.

In virtue of Statement 4.1, there is such a simplification $\widetilde{\Gamma}$ of Γ that among graphs obtained from $\widetilde{\Gamma}$ by one elementary simplification one can choose such a pair Γ_1 and Γ_1' for which one of the graphs Γ_0 or Γ_0' is a minimal representative for Γ_1 but not for Γ_1'. Without loss of generality we shall assume that this graph is Γ_0. According to Statement 4.2, there exists a graph Γ_2' which can be obtained by an elementary simplification from each of Γ_1 and Γ_1'.

Since by assumption Γ_1' does not have Γ_0 among minimal representatives, the graph Γ_2' does not have Γ_0 among minimal representatives. On the other hand, since Γ_0 is a minimal representative for Γ_1, then Γ_1 has at least two minimal representatives.

Changing the notation from Γ_1 to Γ and repeating the above argument, we see that one of the graphs (denote it by Γ_2) obtained by an elementary transformation from Γ_1 has at least two non-isomorphic minimal representatives as well. Arguing as above, we shall get a chain $\Gamma \to \Gamma_1 \to \Gamma_2 \to \ldots$ of graphs, each of which is obtained from the previous one by an elementary simplification, and each Γ_i has at least two minimal representative.

This leads us to the contradiction for that graph Γ_i which is not simplifiable. $\qquad\square$

Thus, we have completely described how to recognise the equivalence of framed 4-graphs as elements of $\mathfrak{G}$: one should take their minimal representatives and compare them. In the case of $\mathfrak{G}$, minimal representatives cannot be simplifiable because they (by definition of minimality) contain no bigons to be canceled by a second Reidemeister move, and the number of unicursal components is equal to one.

In the case of graphs from $\widetilde{\mathfrak{G}}$, a minimal representative can be simplifiable in the case it has a split component. In this case the graph is equivalent to zero.

Analogously to Lemma 4.6 for $\widetilde{\mathfrak{G}}$ one proves the following

Lemma 4.7. *Framed graphs Γ and Γ' from $\widetilde{\mathfrak{G}}$ are equivalent if their minimal representatives $\widetilde{\Gamma}$ and $\widetilde{\Gamma}'$ either both contain split trivial components, or do not contain split trivial components and are isomorphic.*

4.3.3 Application of invariants

Using the parity constructed in Sec. 2.3.3 and the invariant [·] defined for this parity, one can prove the following

Theorem 4.10. *Let K be a framed 4-graph whose Gauss diagram is irreducibly odd. Then every framed 4-graph K' representing the same free knot as K there exists a smoothing which is isomorphic to the framed 4-graph K (as a framed 4-graph).*

Evidently, this theorem yields the following theorem.

Theorem 4.11. *An odd irreducible Gauss diagram D is minimal, i.e., any Gauss diagram D' representing the free knot generated by D has more crossings than D does.*

Example 4.3. A framed 4-graph given by a chord diagram shown in Fig. 4.30, is irreducibly odd.

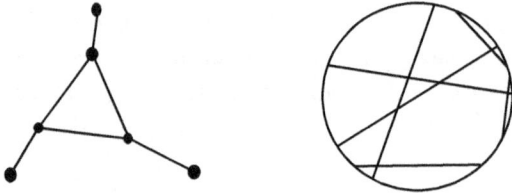

Fig. 4.30   Irreducibly odd chord diagram and its intersection graph

It is easy to see that the set of *irreducibly odd Gauss diagrams* is infinite, thus, the set of different minimal framed 4-graphs is infinite as well.

Theorem 4.10 is a corollary from the following fundamental statement which is easy to prove.

Statement 4.3. *For an irreducibly odd 4-graph K with one unicursal component the following equality takes place:*

$$[K] = K. \tag{4.3}$$

In the left hand side of (4.3), the graph K is considered as a representative of a *free knot*, and in the right hand side it is considered as an element from $\mathfrak{G}$.

The proof of Statement 4.3 is obvious: since all crossings of the framed 4-graph K are *odd*, then in the formula for $[K]$ one has to smooth only the empty set of crossings. Thus, we get the only summand which is the 4-graph K itself.

Proof of Theorem **4.10.** Let us deduce Theorem 4.10 from Statement 4.3. Let a framed 4-graph K' be equivalent to a framed 4-graph K as a free knot. Then $[K'] = [K] = K$. Consequently, the graph K' has at least one smoothing $\widetilde{K}$, which is equivalent to the graph K as an element from $\mathfrak{G}$. Thus K is a minimal representative for $\widetilde{K}$. Furthermore, note that if a framed 4-graph L' is obtained from L by one elementary simplification, then the graph L' is obtained from L by a smoothing in two vertices (just those vertices where the elementary simplification was performed). Consequently, the graph $\widetilde{K}$ is obtained by smoothings of some vertices of the graph K', and the graph K is obtained by smoothing of $\widetilde{K}$. Thus, K is obtained by smoothing from $\widetilde{K}$. $\qquad\square$

Analogous results about minimality can be obtained for links with arbitrarily many components. To that end, instead of the bracket $[\cdot]$ and the parity from Sec. 2.3.1 one can use, for example, the bracket $\{\cdot\}$ and the parity from Sec. 2.3.2. Namely, analogously to Theorem 4.10 one can prove the following theorem.

Theorem 4.12. *Let L be a free two-component link with all crossings belonging to both components such that no second decreasing Reidemeister move is applicable to the framed 4-graph L. Then for every framed 4-graph L' generating the same free link as L, there exists a smoothing which is isomorphic to L as a framed 4-graph.*

Proof. By the definition of the invariant $\{\cdot\}$ for the parity from Sec. 2.3.2, we have $\{L\} = L$ (note that the invariant $[\cdot]$ for the parity from Sec. 2.3.1 does not work, since after smoothing the empty set of crossings of the diagram L having more than one unicursal component, we get L itself, consequently, $[L] = 0$).

Since the graph L cannot be decreased by a second Reidemeister move, we see in $\mathfrak{G}$ that $L \neq 0$. In addition, for every diagram L' which gives the same link as L one has $\{L'\} = L$. Thus, at least one of the smoothings of

L' represents an element equivalent to L in $\widetilde{\mathfrak{G}}$, which, in turn, yields that the framed 4-graph L is a smoothing of L'. □

As an example we give the following

Statement 4.4. *A free link diagram L_1 given in Fig. 4.31, is minimal (with respect to the number of crossings), and the corresponding atom is orientable.*

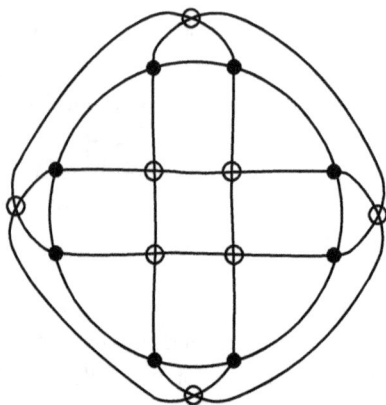

Fig. 4.31   A minimal representative of a two-component link

Proof. The orientability of the corresponding atom follows from a straightforward check of the source–sink condition. The minimality follows from Theorem 4.12. □

Note that examples of minimal diagrams of framed links in the case of non-orientable atom can be obtained in a much easier way: for such an example one may take the simplest two-component framed link with one vertex belonging to both components.

However, the atoms corresponding to this free link as a frame are *non-orientable* since the frame has no source–sink orientation.

It turns out that the methods given above allow one to prove minimality for diagrams of free knots with orientable atoms as well. Further, we shall show how, by using it, one can get non-triviality of *free knots* with orientable atoms, i.e., such knot diagrams given by framed 4-graphs having all vertices being even.

4.3.4 *Non-invertibility of free links*

There is a natural operation of orientation reversal on the set of free knots. Let us introduce the notation: for an oriented free knot K (framed 4-graph) by $\overline{K}$ we shall denote the oriented free knot (framed 4-graph) obtained from K by inverting the orientation. Analogously, for a free link L by $\overline{L}$ we shall denote the oriented free link obtained by inverting the orientation of all components of L.

In the present subsection we shall show the existence of non-invertible free links: $L \neq \overline{L}$. To this end, we shall modify the bracket $\{\cdot\}$ in a way such that those graphs which appear as summands of it, carry the information about the orientation of the initial link. By $|K|, |L|$ we shall denote *unoriented* knots and links obtained from K and L by *forgetting the orientation*.

We shall prove the following

Theorem 4.13. *Let $L = L_1 \cup L_2$ be a framed 4-graph representing a diagram of an oriented free two-component link, and assume the following conditions hold.*

1) *All crossings of L are formed by both components L_1 and L_2.*
2) *There is no room to apply any second decreasing Reidemeister move to L.*
3) *The total number of crossings of L is odd.*
4) *The framed oriented 4-graph L is not isomorphic to any of the graphs $\overline{L}_1 \cup L_2$, $\overline{L}_1 \cup \overline{L}_2$ with respect to the orientation. In other words, the orientation change of the component L_1 does change the 4-graph regardless the orientation of the component L_2.*
5) *There is no isomorphism of the framed 4-graph $|L|$ onto itself taking $|L_1|$ to $|L_2|$ and taking $|L_2|$ to $|L_1|$.*

Then the diagram L is not equivalent to $\overline{L}$.

The first two conditions of the theorem guarantee the minimality of the diagram L according to Theorem 4.12.

It is clear that the diagram $\overline{L}$ is also minimal by the same theorem. Nevertheless, the bracket $\{\cdot\}$ does not allow to distinguish between L and $\overline{L}$ because all summands in the bracket are equivalence classes of *unoriented* framed 4-graphs.

Let us now modify the bracket $\{\cdot\}$. Note that for two-component links the *oddness* of the number of crossings formed by both components, is an invariant property.

Let $\mathfrak{H}$ be the set of equivalence classes of $\mathbb{Z}_2$–linear combinations of framed 4-graphs with two unicursal components, one of which is oriented, by the following two equivalence relations:

1) the second Reidemeister move (taking into account the orientation of the oriented component);
2) the relation $L\sqcup\bigcirc = 0$, where diagrams having a split trivial component, are equated to zero.

For the set of two-component free links with oddly many crossings between components and with *one oriented component*, we shall construct an invariant $L \mapsto \{L\}_2 \in \mathfrak{H}$.

If both components of the link L are oriented and the number of crossings between the components is odd, then there will be two invariants corresponding to L: $\{L\}_{2,L_1}$, $\{L\}_{2,L_2}$ depending on the choice of the oriented component.

Now, let us construct the invariant $\{\cdot\}_2$.

For the two-component link L an *odd* crossing is the crossing formed by both components. Let us consider even smoothings of the diagram L. Each of them will be a framed 4-graph representing a link of at least two components: since we smooth even crossings only, we shall get some set of components originating from L_1 and some set of components originating from L_2.

We shall select only those summands where the number of components is equal to two; besides, we shall endow the component coming from L_1 with an orientation.

The orientation is chosen according to the following rule. For each diagram L_s obtained from L by smoothing of all even crossings, at each odd crossing we have the orientation of the component $(L_s)_1$, corresponding to the orientation of the component L_1 at the same crossing.

The number of such crossings is odd, and in each of them, L_1 generates one of the two orientations of the component $(L_s)_1$. For the orientation of $(L_s)_1$ we choose that one orientation (of two) which occur oddly many times.

More precisely, we have

$$\{L\}_2 = \sum_{s_{2,even}} ((L_s)_{1,odd\ or.}, (L_s)_2) \in \mathfrak{H}.$$

Then the following theorem holds

Theorem 4.14. *The bracket* $\{L\}_2$ *is an invariant of two-component free links with one fixed components and the odd number of mixed crossings between the two components.*

Proof. Let us repeat the proof of Theorem 4.8 paying more attention to the orientation and parity arguments.

First of all, note that the "non-orientable" version $\{\overline{L}\}_2$ of the invariant $\{L\}_2$ is obtained by a natural projection of the invariant $\{L\}$ to the linear space of framed 4-valent graphs generating two-component links.

Thus, it suffices for us to take care about the behavior of orientations of components under Reidemeister moves for those links which appear in $\{L\}$.

Under the first Reidemeister move, the check is evident: the crossing participating in the move is smoothed in a way compatible wit the orientation and it does not affect the resulting orientation of the summand in the bracket.

The same happens in the case of the second Reidemeister move with two even crossings and in the case of the third Reidemeister move with three even crossings: the corresponding diagrams before and after the Reidemeister move have *the same set of odd crossings with the same orientation of the component L_1 in each of them.*

In the case of the second increasing Reidemeister move $L \mapsto L'$, applied to two odd crossings, the summands in $\{L\}_2$ and in $\{L'\}_2$ are in one-to-one correspondence with each other and are obtained from each other by a second Reidemeister move. It remains to show that in every summands of $\{L'\}_2$ both new crossings generate the same orientation of the oriented component. Thus, the existence of these two crossings does not affect the rule for defining the orientation of the oriented component.

Now, let us consider the case of the third Reidemeister move where both components take place. If two of three branches belong to the non-oriented component L_2 and one branch belongs to the oriented component L_1, then both in the diagram L before applying the Reidemeister move, and in the diagram L' after the Reidemeister move we have two consecutive crossings of the same orientation. Consequently, the orientations of the corresponding summands in $\{L\}_2$ in $\mathfrak{H}$ coincide.

We are left with the most difficult case, when the oriented component forms two branches, and the non-oriented component forms one branch, moreover, the only even crossing taking part in the third Reidemeister move, lies on the self-intersection of the oriented component.

The two versions of this move are shown in Fig. 4.32.

In the upper part of Fig. 4.32 we see that the summands from the first pair in the right side and the summands in the left side have two crossings each, moreover, the orientations of the component L_1 in these crossings are *opposite* in both cases, consequently, the resulting orientation for the summand as an element of $\mathfrak{H}$ is the same (up to the second Reidemeister move).

As for the second summand in the upper part of Fig. 4.32, the two intersection points (left and right) of the component L_1 with the component L_2 generate the same orientations as the two intersection points in the right part, thus, in this case the rule for orienting the component for $\{L\}_2$ in the right part is the same as the rule for the left part.

In the lower picture, the first summand in the left part has two crossings with similar orientation, thus, their common contribution to the definition of orientation of L_1 cancels. The same happens in the first summand in the right part.

In the lower picture, the second summand in the left part has two crossing points a, b between the components L_1, L_2, and the second summand in the right part has two crossing points a', b' between the corresponding components. Note that the orientation of the component L_1 in the point a is opposite to the orientation in a', and the orientation in b is opposite to the orientation in b'. If the orientations of the component L_1 in the points a and b agree in the left hand side, then the corresponding orientations agree in the right hand side. If the orientations in a and b in the left hand side are different, then both in the right hand side and the left hand side

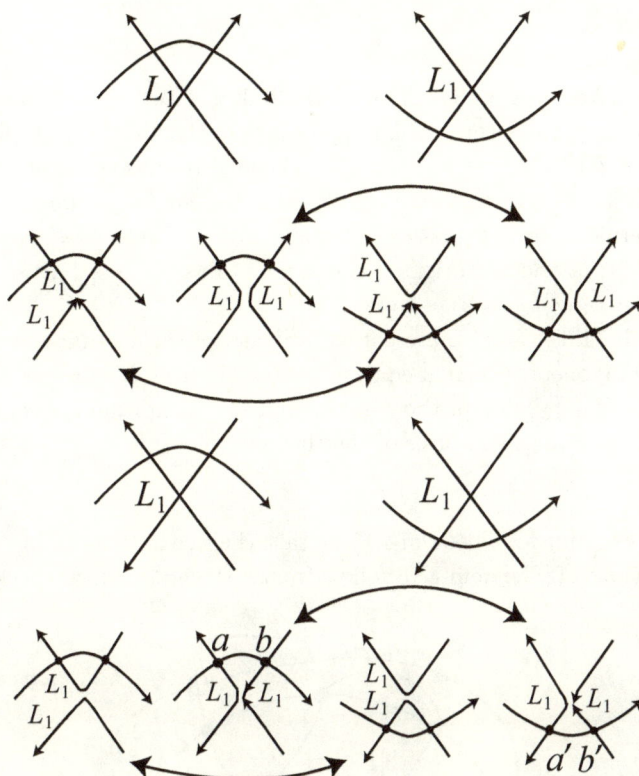

Fig. 4.32   The behavior of orientations under the third Reidemeister move

we have a couple of crossings which give both possible orientations for the components L_1. Summarising the above statements, we conclude that in all these cases the orientation of the corresponding summand in $\{L\}_2$ will be the same.

Thus we have proved that the bracket $\{L\}_2$ is invariant under Reidemeister moves as an element of $\mathfrak{H}$.   □

From the definition one easily gets the following

Statement 4.5. *If for a link diagram $L = L_1 \cup |L_2|$ all crossings belong to both components, and the number of components is odd, then $\{L_1 \cup |L_2|\}_2 =$*

$L_1 \cup |L_2|$ *in* $\mathfrak{H}$.

Proof of Theorem 4.13. Assume the link L satisfies the conditions of the theorem. Assume the link L is equivalent to the link $\overline{L}_1 \cup \overline{L}_2$. Applying the bracket $\{\cdot\}$ to $|L|$, we see that the Reidemeister moves cannot take the link $|L|$ to itself and switch the components L_1 and L_2 (property 5).

Furthermore, $L_1 \cup |L_2|$ is not equivalent to $\overline{L}_1 \cup |L_2|$ as a two-component free link with a selected oriented component. Passing to $\{\cdot\}_2$, we see that $L_1 \cup |L_2| \neq \overline{L}_1 \cup |L_2|$ in $\mathfrak{H}$.

Thus the links L and $\overline{L}$ are not equivalent: for the coordinated enumeration of components this non-equivalence follows from the properties of the new invariant $\{\cdot\}_2$, and for the non-coordinated orientation the equivalence is forbidden by the hypothesis of the theorem. $\square$

As an example for Theorem 4.13 we take the free link shown in Fig. 4.33. All conditions of Theorem 4.13 follow from a straightforward check.

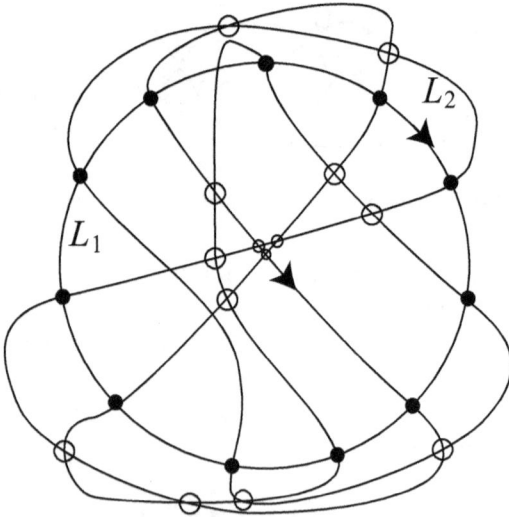

Fig. 4.33   A non-invertible free link

4.4 Goldman's bracket and Turaev's delta

It turns out that on the set of pairs [a closed oriented 2-surface, a curve on this surface] there are interesting operations which are invariant under Reidemeister moves. Namely let S be a fixed 2-surface (we assume this surface to be oriented; otherwise, all arguments below work for the case of coefficients from $\mathbb{Z}_2$).

We think of all curves to be generically immersed. Such a curve represents an embedding of a framed 4-graph. Two immersed curves in general position are *homotopic* if and only if one of them is obtained from the other by a composition of the Reidemeister moves (certainly, in the case when graphs are embedded in a surface, Reidemeister moves respect the structure of this surface, unlike Reidemeister moves for free graphs).

Let Γ_S be a set of all linear combinations of homotopy classes (note that homotopy may violate the smoothness condition whenever a first Reidemeister move is applied) of curves on a fixed surface S with coefficients from the ground field F (the field can be arbitrary in the case of orientable surface S; in the non-orientable case it should be the field of characteristic 2). All curves are assumed to be oriented.

Furthermore, let Γ_S^2 be the set of F-linear combinations of homotopy classes for ordered pairs of oriented curves on S, and let $\Gamma_{S,0}^2$ be the quotient space of the space Γ_S^2 modulo the following relation: $K \sqcup \bigcirc = 0$, i.e., we take to zero all those links having a diagram on S with two connected components, one of which is homotopy trivial.

Now let us pass to the construction of invariant maps.

4.4.1 *The map* $m\colon \Gamma_S^2 \to \Gamma_S$

Let γ_1, γ_2 be a pair of oriented curves generically immersed in S, and let $X_1, \ldots, X_N$ be crossings from $\gamma_1 \cap \gamma_2$ (since the curves are in general position, they are transverse, and there are finitely many of crossings). Let $m(\gamma_1, \gamma_2)_k$ be the oriented curve obtained by smoothing $\gamma_1 \cup \gamma_2$ at X_k in the way compatible with the orientation. We set

$$m(\gamma_1, \gamma_2) = \sum_k \text{sign}_k(1,2) m(\gamma_1, \gamma_2)_k \in \Gamma_S, \qquad (4.4)$$

where the sum is taken over all numbers of crossings $k = 1, \ldots, N$, and $\text{sign}_k(1, 2)$ denotes the sign of the crossing X_k, i.e., it is equal to one if the basis formed by tangent vectors $(\dot{\gamma}_1, \dot{\gamma}_2)$, is positive, and to -1, otherwise.

This sum is considered as an element from Γ_S. A direct check shows that the following theorem holds

Theorem 4.15 (see [20]). *The map $m \colon \Gamma_S^2 \to \Gamma_S$ is well defined, i.e., if the pair (γ_1, γ_2) is equivalent to a pair (γ_1', γ_2'), then $m(\gamma_1, \gamma_2) = m(\gamma_1', \gamma_2')$.*

Remark 4.9. In the case of framed links, the operation (4.4) is defined only over the field of characteristic 2, because we can define the sign $\text{sign}_k(1, 2)$ only if the curves locally lie on an oriented surface; this is the case, say, for flat virtual knots but not for free knots. Moreover, without such a surface, we are unable to define the map m for any *pair of curves*; it is possible to do that only for a two-component free link (without any surface we do not have any intersection points for two curves defined abstractly).

4.4.2 Goldman's Lie algebra

The map m can be treated in another way. Having a pair of curves γ_1, γ_2 (in fact, we talked about a pair of homotopy classes of curves on S), we may think that there is a well-defined map

$$[\cdot, \cdot] \colon \gamma_1, \gamma_2 \mapsto [\gamma_1, \gamma_2] = \sum_k \text{sign}_k(1, 2) m(\gamma_1, \gamma_2)_k,$$

herewith, the permutation of γ_1 and γ_2 leads to the sign change: $[\gamma_1, \gamma_2] = -[\gamma_2, \gamma_1]$.

It can be also easily checked that the operation $[\cdot, \cdot]$ satisfies the Jacobi identity: for any triple of curves $\gamma_1, \gamma_2, \gamma_3$, we have:

$$[[\gamma_1, \gamma_2], \gamma_3] + [[\gamma_2, \gamma_3], \gamma_1] + [[\gamma_3, \gamma_1], \gamma_2] = 0.$$

Thus, the set Γ_S possesses the structure of a Lie algebra [20].

4.4.3 Turaev's delta

Let γ be an oriented curve in general position on an oriented surface S, and let $X_1, \ldots, X_N$ be the intersection points of the curve in S. Then, as a result of smoothing γ at X_i in the way compatible with the orientation, we get two curves, one of which, $\gamma_{i,L}$, can be naturally called *the left one*, and the other one ($\gamma_{i,R}$) is called *the right one*. Thus, we can define a map

$$\gamma \mapsto \sum_i (\gamma_{i,L} \otimes \gamma_{i,R} - \gamma_{i,R} \otimes \gamma_{i,L}). \tag{4.5}$$

This map is skew-symmetrical, however, it is not quite well defined: if we apply the first increasing Reidemeister move to the curve γ, then in the right part of the equality (4.5) we shall have additional summands of the form $\gamma_0 \otimes \gamma' - \gamma' \otimes \gamma_0$, where γ_0 is a contractible curve, and γ' is a curve homotopic to γ.

Thus, in order to have a well-defined cobracket, we have to take the quotient of the set of curves by the relation taking the contractible curve to zero (in the tensor product, we set $0 \otimes a = a \otimes 0 = 0$). The same arguments allow one to construct an invariant map

curve $\mapsto$ linear combination of pairs of curves,

which does not lead to a coalgebra structure, but it suffices to consider a map given by the formula

$$\Delta \colon \gamma \mapsto \sum_s \gamma_{X_s} \in \Gamma_{S,0}^2,$$

where the sum is taken over $\mathbb{Z}_2$, and γ_{X_s} means a pair of curves on the surface S obtained by smoothing the curve γ in the crossing X_S coordinated with the orientation, and the result of the map Δ is considered as an element from $\Gamma_{S,0}^2$.

Thus we have got a map $\Delta \colon \Gamma_S \to \Gamma_{S,0}^2$, to be called *Turaev's delta* in the sequel [65].

Let Γ_{Fr} be the set of all linear combinations of all free knots with coefficients from $\mathbb{Z}_2$, and let $\Gamma_{Fr,0}^2$ be the set of $\mathbb{Z}_2$–linear combinations of two-component free links modulo the following relation: $K \sqcup \bigcirc = 0$, i.e., we take to zero all those free links having a diagram with two components one of which is a split unknot.

Analogously, let us define Turaev's delta from Γ_{Fr} to $\Gamma_{Fr,0}^2$: with each framed 4-graph K with one unicursal component we associate a linear combination $\sum_i K_i$ (over $\mathbb{Z}_2$) of framed 4-graphs with two unicursal components each, obtained by smoothing at corresponding vertices, and the resulting sum will be considered as an element from $\Gamma_{Fr,0}^2$.

4.4.4 *Applications of Turaev's delta*

In Sec. 4.3.3 we have shown that the parity (in sense of Sec. 2.3.1 or in sense of Sec. 2.3.2) can be used for proving minimality of free knot diagrams. However, the very first example (non-triviality of an irreducibly odd knot) deals only with free knots whose Gauss diagrams have all odd chords. We have used the invariant $[\cdot]$ for proving its minimality.

Then we considered an example of a two-component free link whose atom is orientable. By means of the invariant $\{\cdot\}$, applied for the parity from Sec. 2.3.2, we proved minimality of this diagram L *in the strong sense*: we proved that every framed 4-graph L' realising the same link as L, admitted a smoothing which represents a graph isomorphic to L as a framed graph.

Now, we are going to give an example of a *free knot* with *orientable* atom for which one can prove minimality of one of its diagrams by using the above methods.

Statement 4.6. *The diagram K_1 of a free knot shown in Fig. 4.34, is minimal.*

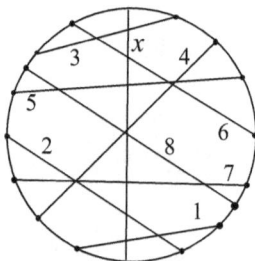

Fig. 4.34 A minimal Gauss diagram

The orientability of atoms corresponding to the diagram K_1 can be easily checked by a straightforward construction of a source–sink orientation.

Proof of Statement 4.6. Let us consider $\Delta(K_1) \in \Gamma^2_{Fr,0}$.

By construction, $\Delta(K_1)$ consists of nine summands (since K_1 has nine vertices), and every summand is a two-component link. These summands are obtained as smoothings of K_1 at crossings. One of these summands (obtained by smoothing along the chord x) is the two-component link L_1, shown in Fig. 4.31. Denote the remaining summands by M_i, $i = 1, \ldots, 8$.

Thus,

$$\Delta(K_1) = L_1 + \sum_i M_i = L_1$$

(in the last equality we used the central symmetry of the chord diagram; for example, the summand M_1 coincides with M_3, etc.).

Considering $\{\Delta(K_1)\}$ and taking into account the invariance of the map $\{\cdot\}$, we see that every diagram of the free knot has at lease one smoothing representing L_1 as a framed 4-graph, i.e., every diagram contains at least 9 crossings. □

Using non-triviality of free links, one can prove non-triviality of free knots. Analogously, non-invertibility of free links yields non-invertibility of free knots. To this end, it suffices to note that Turaev's Δ is orientation sensitive. Moreover, the following theorem holds.

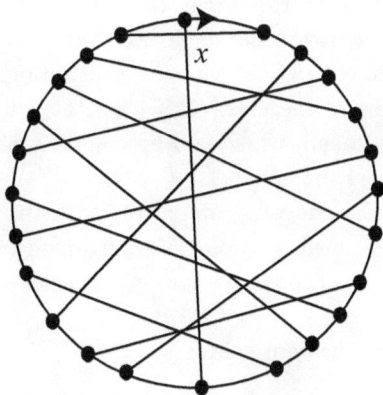

Fig. 4.35   An example of a non-invertible free knot

Theorem 4.16. *The free knot K_2 shown in Fig. 4.35, is not invertible.*

Proof. One can easily see that $\Delta(K_2) = L + \sum_i L_i$, where L is the free link shown in Fig. 4.33, and L_i are two-component free links, each of which has at least one crossing formed by two branches of the same component.

Indeed, for the chord diagram shown in Fig. 4.35, there is exactly one chord x which is linked with all the other chords.

From this we easily see that $\{\Delta(K_2)\}_2 \neq \{\Delta(\overline{K}_2)\}_2$. $\hfill\square$

4.4.5   *Even and odd analogues of Goldman's bracket and Turaev's delta*

Assume in some knot theory a parity is given (usually we shall deal with free knots and the parity from Sec. 2.3.1).

Consider the sets Γ_{Fr} and $\Gamma^2_{Fr,0}$. For them, let us define the maps Δ_{even} and $\Delta_{odd} \colon \Gamma_{Fr} \to \Gamma^2_{Fr,0}$.

Let K be a framed 4-graph with a unique unicursal component. We set

$$\Delta_{even}(K) = \sum_{s=\{v,\ even\}} K_s, \qquad \Delta_{odd}(K) = \sum_{s=\{v,\ odd\}} K_s.$$

Here, in the first case we take the sum of all framed 4-graphs obtained by a smoothing of one even vertex of K (the sum is taken over all even vertices) in the way compatible with the orientation. In the second case, the sum is taken over all odd vertices, and again we take those smoothings of one vertex of the graph K compatible with the orientation.

When applying any of the maps Δ_{even} and Δ_{odd} to a concrete framed 4-graph K we obtain a sum which can be considered as an element from $\Gamma^2_{Fr,0}$.

Note that the term "the way compatible with the orientation" makes sense for unoriented framed 4-graphs as well: among the two smoothings at the vertex one has to choose the one which leads to a two-component link.

Then the following theorem holds

Theorem 4.17. *The maps Δ_{even} and Δ_{odd} are well defined as maps from Γ_{Fr} to $\Gamma^2_{Fr,0}$.*

Proof.   To prove the claim, note the following. If two framed 4-graphs K and K' differ from each other by a first Reidemeister move, then the following equality $\Delta_{odd}(K) = \Delta_{odd}(K')$ holds termwise: the number of odd crossings of K coincides with the number of odd crossings of K', consequently, there is a one-to-one correspondence between the summands in the expansions for $\Delta_{odd}(K')$ and for $\Delta_{odd}(K)$. The corresponding summands are obtained from each other by an application of one first Reidemeister move. In the case of Δ_{even}, in the expansion for K', the number of summands is one more than the number of summands in the expansion for K. This "additional" summand in K' is a two-component link with a split trivial component. Thus, this summand is trivial in $\Gamma^2_{Fr,0}$. The remaining summands in $\Delta_{even}(K')$ are in one-to-one correspondence with summands from $\Delta_{even}(K)$, moreover, these corresponding summands are isomorphic as framed 4-graphs.

Now, let K' be obtained from K by a second Reidemeister move in such a way that the number of crossings of K' is two more than the number of crossings of K. These two "extra" crossings have the same parity. If both crossings are odd, then in the expansions of $\Delta_{even}(K)$ and $\Delta_{even}(K')$ the number of summands is the same, and all the corresponding summands are obtained from each other by means of a second Reidemeister move. If both crossings are even, then in $\Delta_{even}(K')$ we have three extra summands in comparison with $\Delta_{even}(K)$, but two of these three summands coincide identically, and one summand has a split trivial circle. Thus, all new summands contribute zero. All the remaining terms in $\Delta_{even}(K)$ are in one-to-one correspondence with the terms from $\Delta_{even}(K')$, and all corresponding terms are isomorphic as framed 4-graphs.

Analogously, if we consider the case Δ_{odd}, then in the case of two odd crossings we get two "extra" summands in $\Delta_{odd}(K')$ in comparison with those in $\Delta_{odd}(K)$; these two new summands will cancel; in the case of two even crossings we get a one-to-one correspondence between the terms in the expansion. Thus, $\Delta_{odd}(K) = \Delta_{odd}(K')$.

If K' is obtained from K by a third Reidemeister move, then there is a one-to-one correspondence between crossings of the framed 4-graph K and crossings of the framed 4-graph K'. Moreover, even crossings correspond to even crossings, and odd crossings correspond to odd crossings.

If we smooth the diagrams K and K' at v (and the corresponding crossing, to be denoted by the same letter v), and the crossing v does not take part in the third Reidemeister move, then it is evident that the corresponding framed two-component free links will coincide, since the representing graphs are obtained from each other by "the same" third Reidemeister move, which transforms K into K'.

If this crossing v is one of the three crossings taking part in the third

Reidemeister move, then it is easy to see that the smoothing of the diagram K at the crossing v either coincides with the smoothing of the diagram K' at the crossing v' (corresponding to the crossing v) or differs from it by an application of two second Reidemeister move, one of which decreases the number of crossings by two, and the other one increases the number by two. $\square$

Analogously to the "comultiplying maps" Δ_{even}, Δ_{odd} one can construct the "multiplying maps" m_{even}, m_{odd}, herewith for the target space we can take a larger space than $\Gamma^2_{Fr,0}$: it is not necessary to equate to zero those diagrams having a split trivial component in the case of "multiplication" maps.

Remark 4.10. It would be interesting to consider the compositions of maps Δ, Δ_{even} and Δ_{odd} (and their reasonable modifications) taken in different orders and analyze which links (free links) can be obtained from some concrete knot at concrete steps for given parities.

Remark 4.11. In order to define the map Δ we do not need the over/undercrossing structure (which exists in the case of virtual knots).

4.5   Relative parities of links

Theory of links with several components is essentially richer than knot theory both in classical and virtual cases. For instance, for classical links there is an ample theory of link homotopy invariants [23, 57]; the simplest of them is the linking number; and classical knot invariants can be considerably modified after passing to links with many components.

Virtual knot theory proposed by Kauffman [33] includes classical knot theory. It turns out that for virtual knots with one component one can define analogues of *classical link invariants*. For example, in Sec. 1.4 a "self-linking number" of virtual knots was defined. This effect is possible due to parities and coverings over virtual knots: there is a well-defined map from the set of virtual knots to the set of virtual links.

The first constructions of parities in knot theories were simply established on combinatorial or homological properties. The parities constructed below are based on presence of certain geometric subdiagrams (samples) in the knot under consideration. It means that a crossing is called *odd* or *even* depending on presence in the knot some "subknots" which are well situated with respect to the crossing. These reasonings allow one to make parity finer.

Fig. 4.36 shows schematically a transformation of a link and its components (for example, mutation) which changes some sample $P \mapsto P'$. This transformation does not change algebraic invariants which do not use parity and are based on some count (polynomials, homological intersections etc.). It also does not change many simple parities.

New parities are based on counting of intersections with some fixed *sample P*.

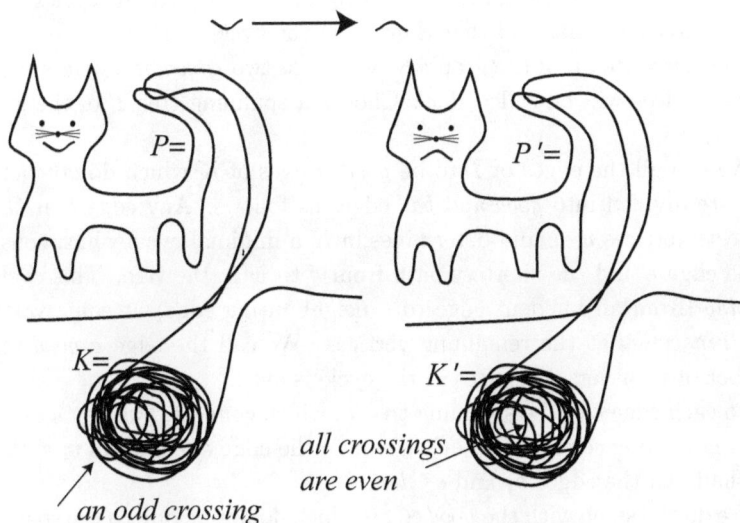

Fig. 4.36 Sample-based parity

Thus, if the sample changes slightly as shown in Fig. 4.36, then the parity changes drastically. For example, the right part of Fig. 4.36 does not contain P at all, so all the crossings are even, whereas some crossings in the left part of the figure are odd.

Let us note that various topological constructions of parity theory lead us to "graphical" results in classical knot theory, see [9, 10, 36].

In this section we define a parity for crossings of one component in two-component link and show how Turaev's delta and the bracket can produce more elaborated parities.

4.5.1   *Projection and twofold covering of link diagram*

Each parity in a knot (link) theory induces two natural maps (projection and a twofold covering) that are constructing in the following way.

Let K be a framed 4-graph. We construct graphs K^2 and K' as follows. Firstly, assume that the graph K is connected.

If K is a cycle, then the graph K' consists of one cycle and the graph K^2 consists of two cycles of the same length. We say that the cycles of K^2 are *dual* to each other and that they *cover* the cycle of K.

To each vertex v of the graph K we assign two *covering* vertices v_1, v_2. These vertices will be called *dual*. Choose a spanning tree T of the graph K.

We call all the edges of T to be *good*. Edges of K which do not belong to T are divided into *good* and *bad* edges as follows. Any edge e in $G \setminus T$ connects vertices v, w and determines in K a minimal cycle which consists of the edge e and the shortest path from v to w in the tree. This cycle is *rotating* (from an incident edge to a neighbouring edge) at some vertices and *transversal* at the remaining vertices. We call the edge e *good* if the number of transversal vertices in the cycle is even.

To each edge e of the spanning tree T, which connects two vertices v, w, we assign two *covering* edges e_1, e_2, where the edge e_i connects v_i with w_i. We shall call the edges e_1 and e_2 *dual*.

We do the same with the *good* edges which do not belong to the spanning tree T.

To each bad edge e, which connects the vertices v and w, we assign two *covering* edges e_1, e_2, the first edge connects v_1 and w_2, the second edge connects v_2 and w_1. The edges e_1, e_2 will be called *dual*.

Let K^2 be the constructed graph. The framing of a vertex of K^2 is naturally induced from the graph K which is covered by K^2. One checks directly that the constructed graph K^2 and the duality relation do not depend on the choice of the spanning tree.

If K is not connected, then the graph K^2 is constructed as the split sum of the graphs K_i^2 that correspond to the connected components K_i of the graph K.

There is a natural involution on the graph K^2 that maps each vertex to the dual vertex and each edge to the dual edge. Since the duality is compatible with the framing relation at a vertex, there is a natural definition of dual components of K^2.

The diagram K' can be obtained from K^2 by removing one of the two sets of components.

The following statement can be easily checked.

Theorem 4.18 (see [50]). *The correspondences $f\colon K \mapsto K'$, $d\colon K \mapsto K^2$ are well defined as a map of links, i.e., Reidemeister moves of the diagram K yield Reidemeister moves of K', K^2.*

Remark 4.12. When K consists of one component, the map $K \mapsto K'$ can be described as follows. The chord diagram $D(K')$ is obtained from the chord diagram $D(K)$ by removing of all the odd chords.

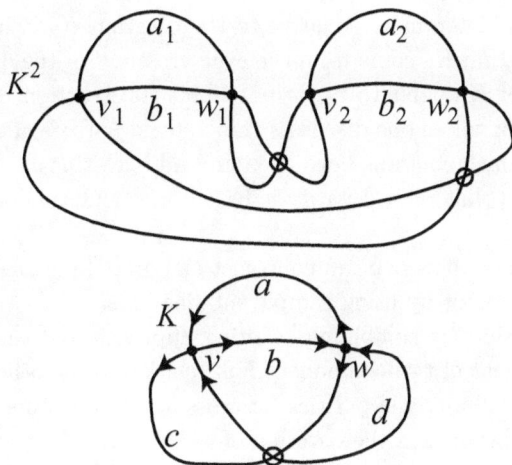

Fig. 4.37   Covering over a framed 4-graph

An example of covering K^2 over a framed 4-graph K is shown in Fig. 4.37.

The graph K drawn below has two vertices v and w. We choose the spanning tree T which consists of the edge b and the vertices v, w. The other edges connect v and w. Among them the edge a is good and the edges c and d are bad. Fig. 4.37 demonstrates the orientation at the vertices v and w compatible along b.

It induces a consistent orientation of the edge a and non-consistent orientation of the edges c and d. The cycle (b, a) is good since it is rotating and the cycles $(b, c), (b, d)$ are bad (each of them is transversal at one vertex).

4.5.2 *Parities relative to a component of a free link*

The first known parity was the Gaussian parity. It was defined on free knots and virtual knots and it is the only non-trivial parity for free knots.

We shall see below that for two-component links one can define a similar parity for one component (for crossings of the other component the parity is not defined).

If we want to combine such a parity with the covering map, we construct for a knot K its covering $K^2 = K_1 \cup K_2$ and get the parity on the crossings of the diagram K_1 relative to the diagram K_2. Since the crossings of the diagram K_2 come from the even crossings of the diagram K the even crossing of K split into two types. This construction was essentially described in [52] where one discusses the "refined parity" of even crossings in a diagram which contains both even and odd crossings.

Nonetheless, that refined parity is based on calculation of intersections in the diagram.

Fig. 4.36 shows how one can construct elaborated parities for one component of a free knot by using sample subdiagrams.

It appears that by combining covering approach and the new parities for one component of two-component link relative to the other component, we can get very non-trivial parities for crossings of a component K_1 in the link $K_1 \cup K_2$ obtained as the covering of a diagram K^2.

Such parities allow one immediately to construct and refine various invariants of the initial diagram K, whose values depend on the presence of one or another sample.

Let $K \cup L$ be a free two-component link that contains even number of mixed crossings.

Let us call a third Reidemeister move of the link $K \cup L$ *special* if it involves one crossing of the component K and two mixed crossings (of the component K with the component L).

Definition 4.8. Let p be a rule which for some class of free links $K \cup L$ assigns a number p_v equal 0 or 1 to each crossing v of the component K. We call p_v *parity for K in $K \cup L$* if the following axioms hold:

a) usual parity axioms (see Definition 2.2) for the Reidemeister moves which involve only crossings of the component K;

b) Reidemeister moves which involve crossings of the component L or mixed crossings do not change parity of the crossings of K (see Fig. 4.38);

c) any special third Reidemeister move does not change the parity of the crossing of the component K.

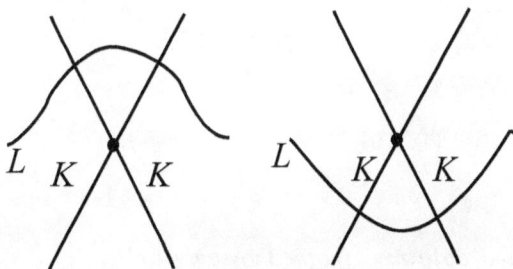

Fig. 4.38   Mixed move does not change the parity

For each crossing v of the component K we define the parity $p_L(v)$ as the parity of number of mixed crossings in a half K_v (the choice of the half does not matter).

Theorem 4.19. p_L *is a parity for K in $K \cup L$.*

The statement follows from the straightforward check of Reidemeister moves.

Example 4.4. Consider the following free two-component link $K \cup L$. The link contains one pure crossing u in the component K and two mixed crossings v and w, one of which lies in one half of K at the crossing u and the other lies in the other half, see Fig. 4.39. By definition the crossing u is odd. The definition implies that any other diagram of the link $K \cup L$ has an odd crossing of the component K, hence it contains mixed crossings. Thus, the considered diagram is minimal.

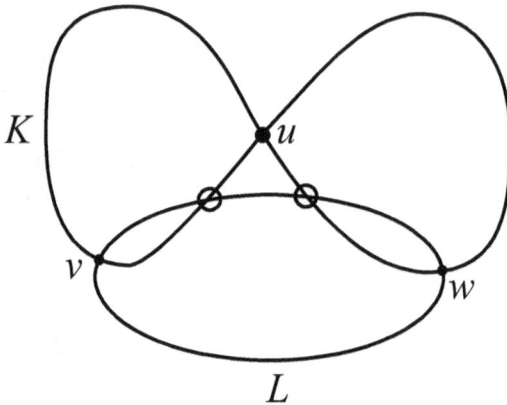

Fig. 4.39   Two-component link $K \cup L$

4.5.3   *Parities obtained from Turaev's delta*

The parity considered above for one component, being relative to the other component, like the Gaussian parity, is defined by an arithmetic calculation of crossings of some type. It appears that one can construct parities which depend on the presence of some "sublink" in the given link. For this we apply some operation (delta or bracket) to the component L of a two-component link $K \cup L$ and look how the result of the operation intersects the component K.

Let us describe the parity construction in more detail.

The first parity will depend on some link $P \cup Q$, the presence of the link as a resolution of the diagram means all the crossings of the diagram are odd.

Let $P \cup Q$ be a free non-splittable two-component link whose component P and Q are not trivial and are not equivalent to each other.

Let $\mathcal{F}$ be the set of equivalence classes of non-splittable free three-component links and let $\mathcal{M} = \mathbb{Z}_2\mathcal{F}$ be the corresponding free module.

We consider a map $f_{P,Q}$ which relates ordered two-component links with an element of the module $\mathcal{M}$. This map is the composition of two maps: $f_{P,Q}(K \cup L) = pr_{.,P,Q} \circ \Delta_L$.

Here Δ_L is the ordered map Δ which maps two-component links to $\mathbb{Z}_2$-linear combinations of three-component links. This map relates a link $K \cup L$ to the sum $\sum_s K \cup L_{s,1} \cup L_{s,2}$, where the resolution $L \mapsto L_{s,1} \cup L_{s,2}$ is taken for all pure crossings s of the diagram L, and we do not consider splittable summands $K \cup L_{s,1} \cup L_{s,2}$; as a result we get a linear combination of three-component links with a distinguished component K.

By applying the projection $pr_{.,P,Q}$, we keep only those summands for which the two-component link obtained from L is equivalent to $P \cup Q$. Since P and Q are not equivalent the components of each summand can be ordered so that the first component is K, the second component is P and the third component is Q.

Statement 4.7. *The map Δ_L is well defined.*

The statement follows from the direct check of Reidemeister moves.

Let us define the parity $p_{P,Q}$ for the knot K relative to L in the link $K \cup L$.

Let $K \cup L$ be a link with even number of mixed crossings. Each pure crossing v of the component K of the link $K \cup L$ corresponds naturally to a crossing in every term $K \cup L_1 \cup L_2$ in the sum $f_{P,Q}(K, L)$. We denote this crossing as v as well. One can consider the two-component sublink $K \cup L_1$ and the parity $p_P(v)$ relative to the component L_1 equivalent to P. Summing these parities for all the summands of the sum $f_{P,Q}(K, L)$, we get the number $p_{P,Q}(v)$.

Statement 4.8. *The parity $p_{P,Q}(v)$ is a well-defined parity for K relative to L.*

Proof. For all Reidemeister moves except for a second move applied to the component L the number of summands in Δ does not change, one can identify all crossings of K in every summand, and the corresponding parities do not change.

A second move on the component L leads to two new summands. Their contributions into the parity for each crossing of the diagram K annihilate, see Fig. 4.40.

The figure shows two diagrams that differ with a second Reidemeister move. Below the two contracting summands are shown.   □

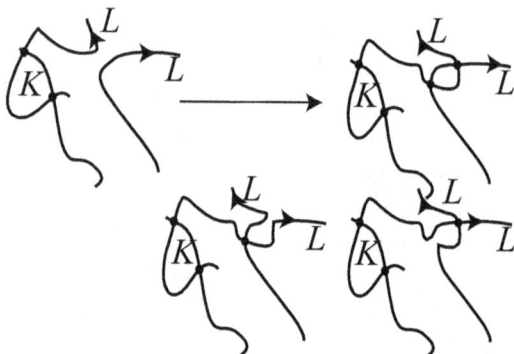

Fig. 4.40   Summands which annihilate by second Reidemeister move

4.5.4   *Parities based on bracket*

The idea of another parity construction is the following. Let $K \cup L$ be a two-component link and let the component L be odd and irreducible. Consider L as a four-valent graph and consider a basis $\alpha_1, \ldots, \alpha_k$ of the homology group $H_1(L, \mathbb{Z}_2)$ of the graph. Each element α_i is presented as a cycle (or sum of cycles) in L. Assume that each of these cycles has even number of intersections with the component K.

Then we can call a crossing v of the component K of the diagram $K \cup L$ to be *even* (resp., *odd*) relative to α_i if the intersection number of a component half at the crossing v with the cycle α_i is even (resp., odd).

The direct check of Reidemeister moves leads to the following result.

Statement 4.9. *The cycles $\alpha_1, \ldots, \alpha_k$ define parities of the knot K relative to L.*

Let us define a linear space $\widetilde{\mathfrak{G}}''$ as the set of $\mathbb{Z}_2$–linear combinations of the following objects. We consider all framed 4-graphs with two ordered unicursal components $K \cup L$ modulo the following equivalence relations:

1) all possible Reidemeister moves for the first component and all possible mixed Reidemeister moves;

2) second Reidemeister moves on the component L.

Let $K \cup L$ be a framed 4-graph with two ordered components K and L. We define an invariant $[K \cup L]_L$ with the formula

$$[K \cup L]_L = \sum_{s_{1,even}} K \cup (L_{s_{1,even}}) \in \widetilde{\mathfrak{G}}'',$$

where the sum embraces all the even resolutions $s_{1,even}$ of the framed 4-graph L, which have one component.

The direct check of Reidemeister moves leads to the following.

Theorem 4.20. $[K \cup L]_L$ *is an invariant of framed links with two ordered components.*

It turns out that this definition can be extended for a parity of the component K in $K \cup L$ by using $[K \cup L]_L$.

4.5.5 *Examples*

Consider the following construction. Let L_1, L_2 be two diagrams of oriented free knots whose edges are marked with points x, y distinct from vertices.

We define the *marked point connected sum* as the free knot obtained by identifying the points x, y (we denote the result point as u). The opposite edges at u are defined according to the orientation so that the resulting framed graph be a knot, see Fig. 4.41. It is necessary to add a virtual crossing in Fig. 4.41 in order to have compatible orientations at the point $x = y$.

Let (L_1, x), (L_2, y) be two minimal knot diagrams with marked points, the knot L_1 be equivalent to P, L_2 be equivalent to Q, and P and Q have the same number k of crossings. Let L be a connected sum with marked points of these diagrams. The knot L has two halves at the crossing u.

Assume that the link $K \cup L$ has one crossing v in the component K and four mixed crossings. Namely, each of halves of the knot K at v has exactly one intersection with each of halves of the knot L at u. Then $p_{P,Q}(v) = 1$. Indeed, the resolution of the component L at every crossing, except u, gives

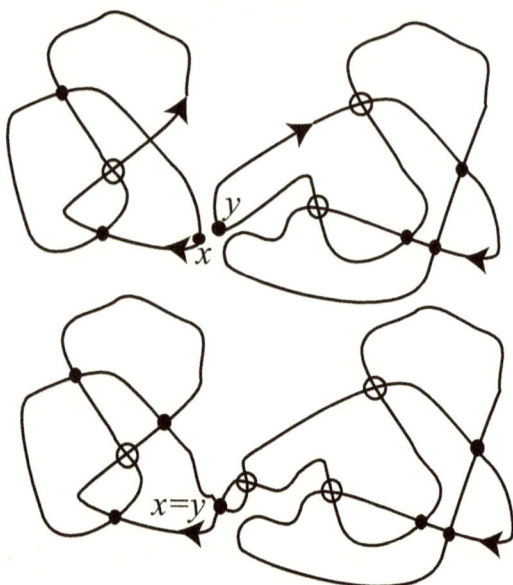

Fig. 4.41   Connected sum of oriented free knots with marked points

one component and the number of its crossings is less than k. Hence, this component is equivalent to neither P nor Q. The smoothing of $K \cup L$ at u gives us $P \cup Q$ by definition. Moreover, after the component Q is removed each of halves of the knot K at u intersects with the component P at even number of points.

Long virtual knots (defined in [45, 56]) have "non-commutativity" phenomenon: connected sums of knots taken in different order are (almost always) not equivalent to each other. The same fact holds for free knots: for different odd irreducible free knots K_1, K_2, K_3, which are not sums of simpler free knots, the diagrams shown in Fig. 4.44 are not equivalent.

Thus, we can take the diagram P, see Fig. 4.44 (left), as the first component of the link $P \cup Q$ and consider the parity $p_{P \cup Q}$. It is clear that the presence of such a sample P depends, for example, on how the summand K_2 is situated in the component L.

Analogous examples can be constructed for a one-component knot K, which splits into a connected sum: by changing the order of summands we

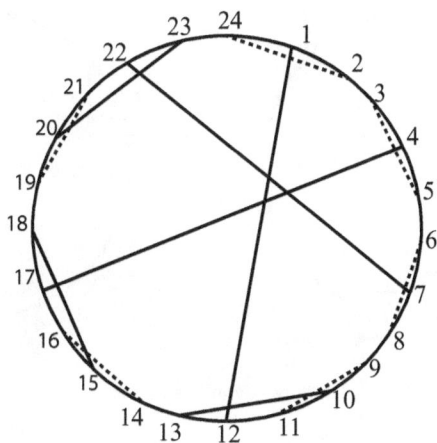

Fig. 4.42   A diagram

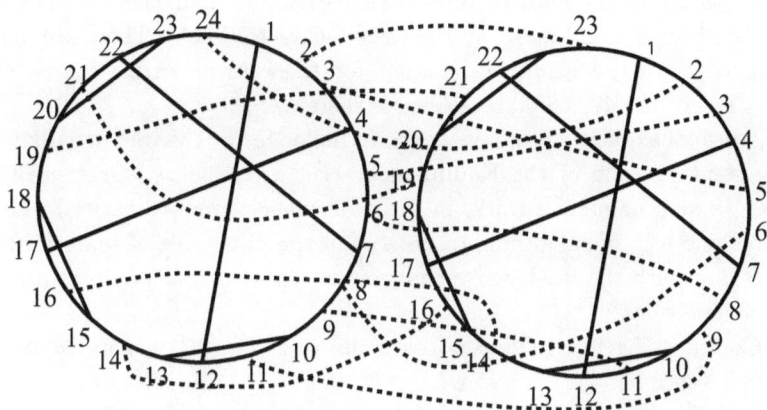

Fig. 4.43   The covering

can change drastically the parities of the covering K^2.

This leads naturally to changing of various invariants based on parities though algebraic and polynomial invariants remain the same. The reason is that parity-based invariants are "pictures" rather than numbers.

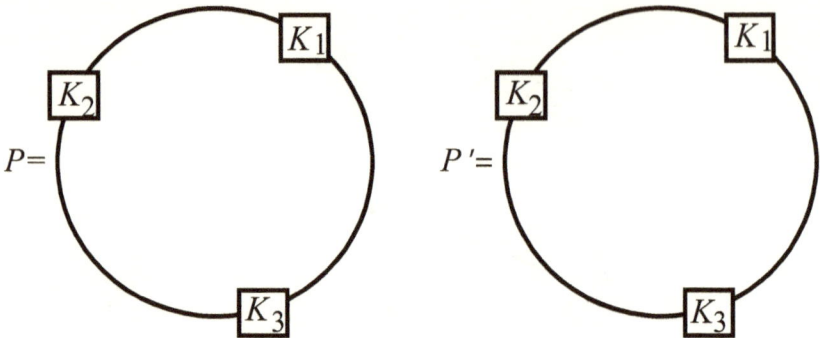

Fig. 4.44   Non-equivalent free knots

4.6   An analogue of the Kauffman bracket

In the present section we are going to construct a refinement of the Kauffman bracket $\langle \cdot \rangle$ for knot theories with parity, this refinement generalises the usual Kauffman bracket in the case of classical knots. There are many refinements of the Kauffman bracket for the case of virtual knots (see, e.g., [2, 11, 12, 47, 58] and references therein).

We shall explicitly write down the formulae in the case of virtual knots. The generalisation of the Kauffman bracket given below works in other cases as well, namely, it works in the cases when there is a natural rule to decree one way of smoothing to be A, and the other rule of smoothing to be B, herewith the Reidemeister moves are "in a natural way coordinated with these rules".

Consider the free module $\mathcal{F}$ over the ring $\mathbb{Z}[a, a^{-1}]$ generated by all framed 4-graphs.

Let $\widetilde{\mathcal{F}}$ be the module obtained by factoring the module $\mathcal{F}$ by the following two relations:

1) the second Reidemeister move,
2) the relation $L \sqcup \bigcirc = (-a^2 - a^{-2})L$, where L denotes any arbitrary framed 4-graph, and $L \sqcup \bigcirc$ is the disjoint sum of L with a split circle.

The algorithmic recognisability of elements from $\widetilde{\mathcal{F}}$ (the existence of a unique minimal representative) is proved analogously to the case of $\widetilde{\mathfrak{G}}$.

Virtual knot theory possesses the Gaussian parity (in the sense of Sec. 2.3.1); for every knot diagram at every classical crossing we shall use the fact that there are smoothings of types A and B (but we are not going to dwell into the *axioms* these crossings, participating in Reidemeister moves, of types A and B have to satisfy).

Now, let us construct the *even Kauffman bracket* valued in the module $\widetilde{\mathcal{F}}$, in the following way:

$$\langle K \rangle_{even} = \sum_{s,\ even} a^{\alpha(s)-\beta(s)} K_s,$$

where $\alpha(s)$ (respectively, $\beta(s)$) is the number of positive $\asymp\ \to\ )($ (respectively, negative $\asymp\ \to\ \smile\frown$) smoothings in the state s, and K_s is the free link diagram obtained by smoothing the diagram K according to the state s, considered as an element from $\widetilde{\mathcal{F}}$.

Remark 4.13. There is a natural map $\widetilde{\mathcal{F}} \to \widetilde{\mathfrak{G}}$ obtained by factoring the quotient field: $\mathbb{Z}[a, a^{-1}] \to \mathbb{Z}_2$, where $a \mapsto 1, 2 \mapsto 0$.

Thus, the invariant $\{\cdot\}$ represents a simplification of the invariant $\langle \cdot \rangle_{even}$.

Theorem 4.21. *The bracket $\langle \cdot \rangle_{even}$ is an invariant of knots from the theory $\mathcal{K}$ with respect to the Reidemeister moves Ω_2, Ω_3. When applying the move Ω_1, the value $\langle \cdot \rangle_{even}$ gets multiplied by $(-a)^{\pm 3}$, and the following normalisation for $\langle \cdot \rangle_{even}$ is invariant under all Reidemeister moves:*

$$X_{even}(K) = (-a)^{-3w(K)} \langle K \rangle_{even},$$

where $w(K)$ stays for the writhe number of the oriented diagram K.

We shall call $X_{even}(K)$ the *even Jones polynomial* (cf. [30]) for the knot K.

Remark 4.14. The even Jones polynomial is a generalisation of the invariant $\{\cdot\}$ for every knot theory with a parity and the rules A and B for smoothings. To get the bracket from the Jones polynomial one takes the map $\mathbb{Z}[a, a^{-1}] \to \mathbb{Z}_2 : a \mapsto 1, 2 \mapsto 0$.

Besides, for the parity in the sense of Sec. 2.3.1 the polynomial $X_{even}(K)$ is a generalisation of the usual Jones polynomial for the case of classical knots and for those knots with corresponding orientable atoms. In this case, in the definition of the bracket $\langle \cdot \rangle_{even}$, all elements K_s are trivial links which in the module $\widetilde{\mathcal{F}}$ are equal to the multiples of the unknot with coefficients equal to some powers of the polynomial $(-a^2 - a^{-2})$. Taking the generator of the module $\mathcal{F}$, generated by the unknot, to be 1, we get the standard Jones polynomial.

Proof of Theorem **4.21.** This proof is completely analogous to the invariance proof for the bracket $\{\cdot\}$. Let K' be obtained from K by one Reidemeister move.

In the case of Ω_1, we have (in $\widetilde{\mathcal{F}}$):

$$\langle \text{⟨⟩} \rangle_{even} = a\langle \text{⟨⟩} \rangle_{even} + a^{-1}\langle \text{⟨⟩} \rangle_{even}$$
$$= (a + a^{-1}(-a^2 - a^{-2}))\langle \text{⟨⟩} \rangle_{even} = (-a^{-3})\langle \text{⟨⟩} \rangle_{even}.$$

Analogously, for the other case of the first Reidemeister move we have:

$$\langle \text{⟨⟩} \rangle_{even} = (-a^3)\langle \text{⟨⟩} \rangle_{even}.$$

When applying a second Reidemeister move we have to distinguish between the following two cases. If both crossings taking part in the second Reidemeister move are odd, then there is a one-to-one correspondence between crossings of the diagrams K and K', which yields a one-to-one correspondence between summands of $\langle K \rangle_{even}$ and $\langle K' \rangle_{even}$. Herewith, the corresponding summands are obtained from each other by applying a second Reidemeister move, thus they represent equal elements from $\widetilde{\mathcal{F}}$.

In the case of two even crossings taking part in the second Reidemeister move, the standard way of proving the invariance for the usual Kauffman bracket under the second Reidemeister move works. As in the standard Kauffman bracket, the three summands in $\widetilde{\mathcal{F}}$ are canceled because of the coefficients a^2, a^{-2} and $(-a^2 - a^{-2})$.

In the case of the third Reidemeister move, one has to distinguish between the cases when the number of even crossings taking part in this move is equal to one or is equal to three.

If the number of even crossings of K (or K') taking part in the third Reidemeister move, is equal to one, then after smoothing the diagrams K and K' at the corresponding crossings v and v' we get the following sums

$$\langle K \rangle_{even} = a\langle K \rangle_{even,+} + a^{-1}\langle K \rangle_{even,-},$$
$$\langle K' \rangle_{even} = a\langle K' \rangle_{even,+} + a^{-1}\langle K' \rangle_{even,-},$$

where the indices plus and minus are responsible for positive (negative) smoothings at the crossings v and v'. It is easy to see that the corresponding summands for K and K' in one of the two smoothings ($+$ or $-$) coincide, and in the other smoothing they differ by an application of the second Reidemeister move to the corresponding graphs, i.e., they are equal in $\widetilde{\mathcal{F}}$.

In the case when all three crossings of the diagram K taking part in the third Reidemeister move $K \mapsto K'$, are even, the proof repeats verbatim the invariance proof for the Kauffman bracket under the third Reidemeister move in the case of classical and virtual knots. $\qquad\square$

Chapter 5

Cobordisms of free knots

5.1 Introduction

A curve immersed in a surface admits a natural notion of *null-cobordance* or *sliceness*: one says that an immersed curve $\gamma \subset S_g$ in an oriented closed 2-surface S_g of genus g is *null-cobordant* (or *slice*) if there exists an oriented 3-manifold M, $\partial M = S_g$, and a disc D properly mapped in M by a smooth map f such that $f(\partial D) = f(D) \cap \partial M = \gamma$.

Analogously, one says that the *slice genus* of $\gamma \subset S_g$ does not exceed h if in the above definition one uses a surface of genus h with one boundary component instead of the disc D.

The first obstructions for curves to be null-cobordant were found by S. Carter, [6]; after that, the theory was also studied by V. G. Turaev [67], K. Orr, and others.

Remark 5.1. In the sequel, we deal only with *generic immersions* of curves in 2-surfaces, unless otherwise is specified. By a generic immersion for a curve we mean an immersion all whose singularities consist of a finite number of self-intersection points and each self-intersection point is a transverse double point.

It can be easily proved that if two curves are homotopic and one of them is null-cobordant, then the other one is also null-cobordant. Thus, one can talk about *null-cobordant classes of homotopy classes of curves*. Moreover, if a curve γ in a surface S_{g+1} does not share a point with a meridian of a handle of the surface, then one can consider the curve γ to lie in the surface S_g obtained from S_{g+1} by cutting S_{g+1} along the curve (meridian) and pasting resulting components of the boundary with discs. It is evident that the pairs (S_g, γ) and (S_{g+1}, γ) are simultaneously either null-cobordant or not.

Therefore, one can talk about *null-cobordant classes of flat virtual knots* [66], which represent equivalence classes of pairs (a circle immersed in an oriented 2-surface, and the surface itself) up to isotopy of curves on our surface and stabilisation/destabilisation.

The paper [47] (full version is published in [50]) pioneered the overall study of *free knots*. One can naturally define a notion of null-cobordant (or slice) for free knots (see below) in such a way that, if a flat knot is null-cobordant, then the free knot corresponding to it, is also null-cobordant.

In the present chapter we extend the notion of parity from one-dimensional objects (curves with self-intersections) to two-dimensional ones (discs with self-intersections), it allows us to construct invariants of slice-ness for free knots. This question (about existence of non-sliced free knots) became actual after the first examples of non-trivial free knots appeared, see accurate definitions of *null-cobordant* free knots and *slice genus* for free knots below, Definition 5.7.

The obstructions for curves to be null-cobordant (and the corresponding construction of cobordism invariants) suggested by J. S. Carter, K. Orr, V. G. Turaev cannot be straightforwardly defined for the case of free knots, since they use some homological data of the surface, which free knots do not possess. In some sense, parity can replace homological/homotopy information, when no "genuine" homology is present.

The concept of parity has some other applications in the cobordism theory for free knots and immersed curves. In particular, if a free knot represented by a framed 4-graph Γ is slice, then so is the free knot represented by a framed 4-graph Γ' obtained from Γ by "killing odd crossings" (see Theorem 5.9 ahead).

The aim of the chapter is to construct one simple (in fact, integer-valued) invariant of free knots which gives an obstruction for a free knot to be slice (null-cobordant).

In [27] we introduced an equivalence relation related to the notion of cobordism which we called *combinatorial cobordism*: instead of a topological definition using the notion of a *spanning disc* we dealt with a formal *combinatorial* definition following [67]. According to this definition two free knots are (*combinatorially*) *cobordant* if one of them can be transformed to the other one by a finite sequence of moves from a given list. These moves include all the Reidemeister moves, and each of them corresponds to a "real" (topological) cobordism.

This chapter is organised as follows. First we consider a notion of combinatorially cobordant free knots. We construct an invariant and example showing that free knots are non-trivial in the wide sense.

Further, we construct an invariant for topologically cobordant free knots. We prove its invariance under the Reidemeister moves. Then, to show that the invariant is well behaved under cobordisms, we have to extend the notion of parity from one free knot to the spanning disc of the cobordism. This is done by marking double lines of the spanning disc as "even" and "odd". We use homological approach to the definition of parity of a double line which agrees with the homological definition of parity of crossings of a free knot.

After that we give basic definitions of Morse theory for cobordisms of immersed curves, and outline the proof of the main theorem. Taking a Morse function on a spanning disc and having the homological definition of parity for double lines, one extends the invariant to all level lines of this function. Non-triviality of the initial invariant coupled with simple Morse theoretic arguments and the "additivity property" of the invariant leads to a contradiction. The key point in the proof is the way to extend the notion of parity from self-intersection points of a curve to double lines of surfaces.

5.2 Combinatorial cobordism of free knots

Let D be a chord diagram.

Definition 5.1. By an *even symmetric configuration* C on a chord diagram D we mean a set of pairwise disjoint arcs C_i on the core circle of the chord diagram which possesses the following properties:

1) the ends of the arcs do not coincide with chord ends, and the number of endpoints of chords inside any arc is even;
2) every chord having one endpoint in C has the other endpoint in C;

3) consider the involution i of the core circle which fixes all points outside the arcs C_i and reflects all arcs along the radii connecting the centre of the core circle with the middle of the arc. Connecting by a chord the images of two points which formed a chord in the initial chord diagram, we get the chord diagram $i(D)$. We require that the configuration C is symmetric, i.e., the chord diagrams D and $i(D)$ are equal.

Definition 5.2. By an *elementary cobordism* we mean a transformation of a chord diagram deleting all chords belonging to an *even* symmetric configuration, as well as the inverse transformation. We say that two chord diagrams are *cobordant* if one can be obtained from the other by a sequence of elementary cobordisms and third Reidemeister moves.

Remark 5.2. The parity of chords does not change under elementary cobordisms.

The definition of cobordisms given above agrees with the definition of *word cobordism* (nanoword cobordism) [67], since there is a natural map from the cobordism classes of (nano)word to the cobordism classes of free knots.

Note that the first two Reidemeister moves are particular cases of elementary cobordisms, unlike the third Reidemeister move. Therefore, it makes sense to talk about *cobordism classes of free knots*.

The main result of this section is a proof of existence of free knots being not combinatorially cobordant to the trivial free knot. To solve this problem, we shall construct a combinatorial cobordism invariant of free knots.

Let L be a framed 4-graph. Each unicursal component L_i of L can be treated as a framed 4-graph with the Gauss diagram D_i. Thus, some vertices of the graph L can be represented by chords of one of D_i's (namely, those vertices lying on one unicursal component). Among these, let us choose *even* vertices (in the sense of the Gauss diagram D_i), and at each even vertex v of L, we consider the smoothing L_v (one of the two possible) for which the number of unicursal components is greater than that of L by 1.

Now let $\mathcal{R}$ be the set of $\mathbb{Z}_2$–linear combinations of equivalence classes of framed 4-graphs modulo the second and third Reidemeister moves. Set

$$\Delta(L) = \sum_v L_v \in \mathcal{R},$$

where the sum is taken over all even crossings v.

Remark 5.3. In this section, we use the same notation Δ as well as for Turaev's delta, since these two operations, in some sense, are similar. In Turaev's delta a linear combination of ordered pairs of curves is assigned to each curve, and, here, we assign a linear combination of equivalence classes of framed 4-graphs with $k+1$ unicursal components to each framed 4-graph with k unicursal components.

Statement 5.1. *The map Δ is a well-defined map from $\mathcal{R}$ to $\mathcal{R}$.*

Proof. Indeed, assume that L is obtained from L' by a third Reidemeister move. Consider the three vertices a_1, a_2, a_3 of L involved in this move, and the corresponding vertices a_1', a_2', a_3' of L'. By construction, the vertex a_i lies in one unicursal component of L if and only if the vertex a_i' lies in one unicursal component of L'. Moreover, a_i is even if and only if a_i' is even. It is now easy to see that whenever a_i is even, the smoothing L_{a_i} gives the same impact to $\mathcal{R}$ as that of $L'_{a_i'}$ (the corresponding framed 4-graphs are either isomorphic or differ by a second Reidemeister move).

Now, let L and L' differ by a second Reidemeister move, and let L' have two more vertices a, b in comparison with L. If both vertices a, b are odd, then the summands in $\Delta(L)$ are in one-to-one correspondence with those in $\Delta(L')$ and the corresponding diagrams in each pair differ by a second Reidemeister move. If both a and b are even, then it is obvious that the smoothings at these crossings give equal impact to L', and since we are working over $\mathbb{Z}_2$, they cancel each other. $\qquad\square$

Therefore, for any $k \in \mathbb{N}$ the map Δ^k, the iteration of k times of the map Δ, is a well-defined map. So, if L and L' are two framed 4-graphs which are obtained from each other by a third Reidemeister move, then for every positive integer k we have $\Delta^k(L') = \Delta^k(L)$.

Let L be a framed 4-graph with k unicursal components. With L we associate a graph $\Gamma(L)$ (not necessarily 4-valent, but without loops and multiple edges) and a number $j(L)$ according to the following rule. The graph $\Gamma(L)$ will have k vertices which are in one-to-one correspondence with unicursal components of L. Two vertices are connected by an edge if and only if the corresponding components share an odd number of points. The following statement is evident.

Statement 5.2. *If two framed 4-graphs L and L' are equivalent, then $\Gamma(L)$ and $\Gamma(L')$ are isomorphic.*

Define the number $j(L)$ *from the graph* $\Gamma(L)$ in the following way. If $\Gamma(L)$ is disconnected, we set $j(L) = 0$. Otherwise, $j(L)$ is set to be the number of edges of $\Gamma(L)$.

Fix a natural number n. Let $\mathcal{N}$ be a linear space generated over $\mathbb{Z}_2$ by formal vectors $\{a_i, i \in \mathbb{N}\}$. For a framed 4-graph L, we set $\mathcal{J}(L) = a_{j(L)}$ if $j(L) > 0$ and $\mathcal{J}(L) = 0$, otherwise. We extend this map to $\mathbb{Z}_2$–linear combinations of framed 4-graphs by linearity.

Now, set $I^{(n)}(L) = \mathcal{J}(\Delta^n(L))$.

Theorem 5.1. *If L and L' are combinatorially cobordant, then $I^{(n)}(L) = I^{(n)}(L')$.*

The proof of this theorem follows from two statements, first of them, Statement 5.2, is evident, and the second one, Statement 5.3, is a central. In virtue of Statement 5.2, the mapping $I^{(n)}(\cdot)$ is invariant with respect to the third Reidemeister move (since so is Δ^n). Moreover, the following holds.

Statement 5.3. *If L_2 is obtained from L_1 by elementary cobordisms, then $I^{(n)}(L_1) = I^{(n)}(L_2)$.*

Having proved Statement 5.3, we shall get Theorem 5.1.

Sketch of the proof of Statement 5.3. Instead of framed 4-graphs we shall consider Gauss diagrams.

Let D_2 be the Gauss diagram obtained from a Gauss diagram D_1 by deleting an even symmetric configuration C.

The chords of D_1 belong to three sets:

1) the set of those chords corresponding to chords of D_2; we denote them for both D_1 and D_2 by γ_j's.

2) the set of those chords β_j which are fixed under the involution i on C;
3) the set of pairs of chords α_k and $\bar{\alpha}_k = i(\alpha_k)$ which are obtained from each other by the involution i (here $\bar{\bar{\alpha}}_k = \alpha_k$).

Recall that for every chord diagram D, $\Delta^n(D)$ is a sum of some consecutive smoothings $\Delta^{(p_1 \cdots p_k)}(D)$ along chords $p_1, \ldots, p_k$, where all p_i's are chords of D and p_i occurs to be even after smoothing all $p_1, \ldots, p_{i-1}$.

$\Delta^n(D_1)$ naturally splits into 3 types of summands:

1) those summands where all chords p_i are some γ_j. These smoothings are in one-to one correspondence with smoothings of D_2. We claim that the corresponding elements $I(\Delta^{(p_1 \cdots p_k)}(D_1))$ and $I(\Delta^{(p_1 \cdots p_k)}(D_2))$ are equal;
2) those summands where at least one of p_i is β_j, and neither α's nor $\bar{\alpha}$'s occur among p_i. We claim that each of these summands $I(\Delta^{(p_1 \cdots p_k)}(D_1))$ is zero;
3) those summands where at least one of p_i's is α_j or $\bar{\alpha}_j$. These summands are naturally paired: the elements $I(\Delta^{(p_1 \cdots p_k)}(D_1))$ and $I(\Delta^{(\bar{p}_1 \cdots \bar{p}_k)}(D_1))$ are equal.

Let us consider the first case. Since arcs of our even symmetric configurations have no common points with chords γ_j, we see that after smoothing along any of γ's, every arc will completely belong to one circle. This means that the corresponding graphs $\Gamma(\Delta^{(p_1 \cdots p_k)}(D_1))$ and $\Gamma(\Delta^{(p_1 \cdots p_k)}(D_2))$ are isomorphic.

Indeed, chords β_j do not change the graph $\Gamma(\Delta^{(p_1 \cdots p_k)}(D_1))$ at all, since they always lie on one unicursal component. Chords α_i and $\bar{\alpha}_i$ belong to either one unicursal component or the same pair of unicursal components. Therefore, they either do not take part or their impacts to the construction of the graph $\Gamma(\Delta^{(p_1 \cdots p_k)}(D_1))$ cancels.

Let us prove the second case. Let us take one chord $p_j = \beta_k$, and consider the arc C_i of D_1 where β_k lies. Without loss of generality, we may assume that β_k is the innermost chord in C_i amongst those chords β_l we use for smoothings.

Now, our summand looks like $\Delta^{(\cdots \beta_k \cdots)}$. Smoothing the corresponding diagram (obtained as a result of smoothings along the chords being before the chord β_k) along β_k cuts the free knot (the unicursal component) which contains the arc C_i. It is obvious that this unicursal component will be split in the sense of the graph Γ: it will give a new vertex corresponding to the unicursal component which shares chords of only types α and $\bar{\alpha}$ with other unicursal components. Since we do not smooth those chords, then this unicursal component will always share an even number of chords with each unicursal component, i.e., the graph is not connected.

The third case is proved by taking account the following fact. We can always realise whether two ends of a chord belong to either two different circles or the same circle, and whether two chords ends of which lie on distinct circles connect either two distinct circles or the same (see Theorem 7.8 in [56]).   □

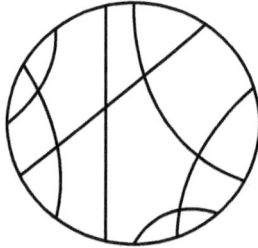

Fig. 5.1   A Gauss diagram of the free knot not cobordant to the unknot

Example 5.1. Consider the free knot K represented by the Gauss diagram shown in Fig 5.1. We have

$$\Delta^3 \left(\vcenter{\hbox{⊗}} \right) = \Delta \left(\vcenter{\hbox{[diagram]}} \right) + \Delta \left(\vcenter{\hbox{[diagram]}} \right) + \Delta \left(\vcenter{\hbox{[diagram]}} \right)$$

$$+ \Delta \left(\vcenter{\hbox{[diagram]}} \right) + \Delta \left(\vcenter{\hbox{[diagram]}} \right) + \Delta \left(\vcenter{\hbox{[diagram]}} \right)$$

$$= \vcenter{\hbox{[diagram]}} + \vcenter{\hbox{[diagram]}} + \vcenter{\hbox{[diagram]}} + \vcenter{\hbox{[diagram]}} + \vcenter{\hbox{[diagram]}} +$$

$$+ \vcenter{\hbox{[diagram]}} + \vcenter{\hbox{[diagram]}} = \vcenter{\hbox{[diagram]}} + \vcenter{\hbox{[diagram]}} .$$

Thus $I^{(3)}(K) = a_4$. Thus, by Theorem 5.1, the cobordism class of K is non-trivial.

5.3 An invariant of free knots

In the present section, we shall construct an invariant of free knots constructed from the parity and justified parity, and prove its invariance. We shall later prove that this invariant delivers a sliceness obstruction for a free knot. Within the present section, by *parity* and *justified parity* we mean the Gaussian parity and the Gaussian justified parity.

An extension of the invariant to be presented below, was constructed in [40]; for our purposes (sliceness obstruction) the version given here will suffice. Nevertheless, it is important to investigate the invariant from [40] (in [40] its invariance under the Reidemeister moves for free knots was proved) from the point of view of sliceness.

To construct the invariant we need to introduce the notion of *justified parity* which is analogous to a parity.

Definition 5.3. By a *justified parity* of crossings we mean a parity where each odd crossing is marked by a letter b or b' (in these cases we call a crossing *an odd crossing of the first type* or *an odd crossing of the second* *type*, respectively), so that the following holds:

1) if a second Reidemeister move is applied to two odd crossings, then they are of the same type (either both are b or both are b');

2) if for a third Reidemeister move we have two odd crossings, then each of them changes its type after the Reidemeister move is applied (the crossing marked by b before the Reidemeister move is applied, should correspond to the crossing marked by b' after the Reidemeister move is applied);

3) moreover, odd crossings not taking part in the Reidemeister move, do not change their types.

We define the *Gaussian justified parity* on framed 4-graphs with one unicursal component, as follows.

Definition 5.4. Let D be a chord diagram with the Gaussian parity, i.e., a chord of D is *even* if and only if the number of chords linked with it, is even, and *odd*, otherwise. Furthermore, an odd chord is *of the first type* (after Gauss) if it is linked with an even number of even chords; otherwise, an odd chord is said to be *of the second type* (after Gauss).

For the framed 4-graph Γ corresponding to a chord diagram D the Gaussian parity and justified parity are defined as those of the corresponding chord diagram.

It can be easily checked that the Gaussian parity and the Gaussian justified parity satisfy the axioms of parity and justified parity.

Definition 5.5. A section of a cobordism, i.e., a section (level line) of the spanning disc, is called *singular* if it contains a critical point, which is neither Morse critical point not Reidemeister singularity. Otherwise, a section is called *regular*.

Later, for cobordism purposes we shall then extend the notion of the Gaussian parity and justified Gaussian parity for another situation. First, we shall define the Gaussian parity and justified Gaussian parity for double lines of a 2-disc with generic intersections, and then for every regular section of this disc (which will be a framed 4-graph representing a free link) we shall define the parity and justified parity for crossings to be the parity and justified parity of double lines it comes from.

However, for the first goal (the construction of an invariant of free knots) it would be sufficient for us to have a well-defined parity for just framed 4-graphs with one unicursal component.

Let us consider the group

$$G = \langle a, b, b' \,|\, a^2 = b^2 = b'^2 = e, \, ab = b'a \rangle$$

with the unit e. Note that G is isomorphic to the infinite dihedral group. For a word γ in the alphabet consisting of a, b, b' we shall denote by $[\gamma]$ the element of G corresponding to γ.

Our first goal is to construct an invariant of free long knots (respectively, of compact free knots) valued in G (in the set of conjugacy classes of the group G).

Let D be an oriented chord diagram with a marked point X on the core circle C distinct from any chord end. Later, we shall see how one can get rid of the orientation of D.

We distinguish between *even* and *odd* chords of D; moreover, we distinguish between *two types of odd chords* of D.

With a marked oriented chord diagram (D, X) we associate a word in the alphabet $\{a, b, b'\}$, as follows. Let us walk along the core circle C of the chord diagram starting from X. Every time we meet a chord end, we write down a letter a, b or b' depending on whether the chord whose endpoint we met, is even, first type odd, or second type odd. Having returned to the point X, we obtain a word $\gamma(D, X)$; this word determines an element of G; by abuse of notation we shall denote this element just by $\gamma(D, X)$. Moreover, sometimes we shall omit X from the notation when it is clear from the context which initial point we have chosen.

Theorem 5.2. *If two marked chord diagrams (D, X) and (D', X') generate equivalent free knots, then $[\gamma(D, X)] = [\gamma(D', X')]$ in G.*

Proof. Indeed, if D and D' differ by a first Reidemeister move (say, D' has one extra chord with respect to D), then the word $\gamma(D')$ is obtained from $\gamma(D)$ by an addition of two consecutive letters $a \cdot a$; thus, the corresponding elements from G coincide.

Analogously, if D' obtained from D by an increasing second Reidemeister move, then the two new chords of D' are of the same parity (and, if they are odd, of the same type); denote the letter corresponding to each of these two chords $(a, b$ or $b')$, by u. Thus, the word $\gamma(D', X')$ is obtained from $\gamma(D, X)$ by addition of $u \cdot u$ in two places. As in the first case, it does not change the corresponding element of G.

The third Reidemeister move $D \to D'$ may be of one of the two types. In the first case, all three chords participating in the third Reidemeister move, are even.

In this case the words $\gamma(D)$ and $\gamma(D')$ coincide identically.

In the second case, two of the three chords taking part in the Reidemeister move, are odd, and one chord is even. Recall that under the third Reidemeister move each of odd chords participating in the move, changes its type.

Consider those three segments of the words $\gamma(D)$ and $\gamma(D')$ where the ends of the three moving chords are located. For those segments containing an end of the odd chord, we get one of the two substitutions $ab \longleftrightarrow b'a$ or $ba \longleftrightarrow ab'$. Both changes correspond to some relations in G.

Now consider the segment of the diagram containing the two ends of the odd chords. If these two odd chords are of the same type in D, then on D' they are of the same type as well. Consequently, when passing from $\gamma(D)$ to $\gamma(D')$ we replace $b \cdot b$ by $b' \cdot b'$ or vice versa. Since both subwords correspond to the trivial element of G, we have $\gamma(D) = \gamma(D')$ in G.

Finally, if the two chords participating in the third Reidemeister move, are of different types on the diagram D, then when passing from $\gamma(D)$ to $\gamma(D')$ the fragment of the corresponding word *stays the same*. Indeed, the adjacent letters b and b' change their position twice because the chords change their type and chords' ends change their positions.

Thus, no Reidemeister move changes the element of G corresponding to the oriented chord diagram with a marked point.   □

This theorem immediately yields the following statement.

Corollary 5.1. *The conjugacy class of the element* $[\gamma(D, X)]$ *in G is an invariant of free knots given by the diagram D, i.e., it does not depend on the marked point X.*

Indeed, moving the marked point through a chord end corresponds to a cyclic permutation of the letters, which, in turn, generates a conjugation in G.

5.3.1   *The Cayley graph of G*

Its Cayley graph looks like a vertical strip on a squared paper between $x = 0$ and $x = 1$: we choose the point $(0,0)$ to be the unit in the group; the multiplication by a on the right is chosen to one step in a horizontal direction (to the right if the first coordinate of the point is equal to zero, and to the left if this first coordinate is equal to one), the multiplication by b is one step upwards if the sum of coordinates is even and one step downwards if this sum is odd, and the multiplication by b' is one step downwards if the sum of coordinates is even and one step upwards if the sum of coordinates is odd (see Fig. 5.2).

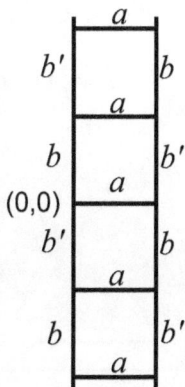

Fig. 5.2   The Cayley graph of the group G

With each pointed chord diagram (D, X) one associates an element from G. Let us show that each element has coordinates $(0, 4m)$.

Consider a chord of type b'. Since by definition there are odd number of ends of even chords and even number of ends of odd chords between the ends of the chord, then before meeting the first end of the chord we have the coordinate (k, l) and before meeting the second end of the chord we have the coordinate (p, q), where k and p are of different parities, and l and q are of the same parity. Hence, the total coordinate shift we have when we pass through the ends of the chord, is $(0, 2)$ (if l and q are odd) or $(0, -2)$ (if l and q are even). Analogously, we can show that total coordinate shift at the ends of a chord of type b is $(0, \pm 2)$. Since the number of chords of type b and b' is even the word in G corresponding to a chord diagram has coordinates $(0, 4m)$.

Moreover, the conjugacy class of the element $(0, 4m)$ for $m \neq 0$ consists of the two elements: $(0, 4m)$ and $(0, -4m)$. Thus, for each *long free knot* one gets an integer-valued invariant, equal to $l = 4m$; we shall denote this invariant for a knot K by $l(K)$; each compact free knot has, in turn, the invariant equal to the absolute value $|l|$; we shall denote the latter by $L(K)$.

It is obvious that if we invert the orientation of the chord diagram, we shall reverse the order of letters in the word γ; this leads to the switch $(0, 4m) \to (0, -4m)$. So, the invariant $L(K)$ can be defined for *unoriented* free knots. The last fact immediately yields two corollaries:

Corollary 5.2. *If for an oriented free knot K we have $l(K) \neq 0$, then K is non-invertible.*

Corollary 5.3. $L(K)$ *is an invariant of* unoriented *free knots.*

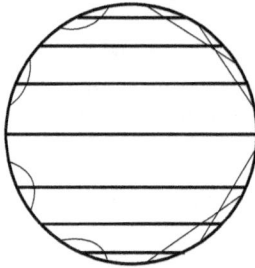

Fig. 5.3   A non-slice free knot

In Fig. 5.3, we have a free knot K_1 for which $L(K_1) = 16$.

We use *bold* lines for describing even chords. The corresponding word in G (with an appropriate choice of the marked point) looks like $(b'a)^7 b' b (ab)^7 = (b'b)^{16}$.

5.3.2 Remarks on the definition of the invariant L for links

Note that Theorem 5.2 works for *any parity*, not only for the Gaussian one, and the proof in the case of any parity follows line by line the proof of Theorem 5.2.

For cobordism purposes, we shall be able to understand the behavior of the invariant L not only under Reidemeister moves, but also under Morse bifurcations. Here we shall only use the Gaussian justified parity.

Moreover, first we have to *define* this invariant for links with many components, since after a Morse bifurcation we can obtain a link from a knot.

Our further strategy is as follows. Assuming we have a cobordism (see Definitions 5.6 and 5.7) $\mathcal{D} \to D$ spanning a framed 4-graph Γ, we shall define the *parity and justified parity* for this cobordism, i.e., we say which of *double lines* (i.e., those lines on D having preimages consisting of two connected components on $\mathcal{D}$) are *even*, and which ones are *odd*, and, furthermore, each odd double line will be divided into segments: *odd of the first type* and *odd of the second type*. This parity will be defined in such a way that the parity (the justified parity) for a double point on Γ will coincide with the parity (justified parity) for the double line, this point belongs to. Besides, this approach will allow us to define the parity and the justified parity for any *generic section* of a cobordism; such a section is a framed 4-graph representing a multicomponent link. With these parity and justified parity, we shall be able to extend our invariant L to sections of L with respect to a Morse function on D and then understand the behavior of this invariant under Morse bifurcations. As will turns out values of these invariant (sets of numbers) will behave nicely under Morse bifurcations, as a result, we get their invariance.

For a single component free link (i.e., a free knot), the value of the invariant γ can be expressed by one non-negative integer L. For a cobordism, every section is a multicomponent free link, so, we have to define the invariant γ on any component to be a collection of conjugacy classes of elements of G (one for each component), and we may require that these elements of G (or the corresponding conjugacy classes) are expressed by one integer (respectively, non-negative integer) number each. In this case we shall be able to associate an integer number (respectively, non-negative integer) with each component of a link appearing in a section of our cobordism.

To this end, we shall need that:

1) the parity and justified parity are well defined for sections and well behaved under Morse bifurcations;

2) the number of intersection points of each unicursal component of a section with each double line is even. Moreover, the number of intersection points of this unicursal component of the section with odd double lines is even; this condition is necessary in order that the element of G corresponding to this unicursal component, has in its representation en even number of the letter a and can be described by one coordinate on the Cayley graph;

3) the value of the first coordinate of the element of G (which is the same as L) behaves well with respect to Morse bifurcations (we shall describe the exact meaning of this below).

In particular, every component of a non-singular level link has an even number of intersection points with double lines: it is necessary for the parity to be well defined. Indeed, in order to define the Gaussian parity of some crossing, one has to take some "half" of the circle corresponding to this crossing and count the number of intersection points belonging to this half. In the case of a free knot, the parity of this number of points does not depend on the half one chooses because it equals the parity of the number of chords linked with the chord in question.

When we have a two-component link, and we take a crossing formed by a single component, the two parities corresponding to the two halves will be different if the total number of crossings between components is odd. So, for those two-component links having an odd number of intersection points between components, there is no immediate way to extend the Gaussian parity.

As we shall see further, all these conditions necessary for naturally extending the Gaussian parity will be automatically satisfied for those multi-component link which are *sections* of a disc cobordism.

In order to do this we need to give a more topological definition of the Gaussian parity.

5.4 Slice genus and cobordisms of free knots

Definition 5.6. Let K be a framed 4-graph with one unicursal component. We say that K has *slice genus at most g* if there exist a surface $\mathcal{D}_g$ of genus g with one boundary component (circle) S, a 2-complex $D_g \supset K$ containing K as a subcomplex, and a continuous map $\nu \colon \mathcal{D}_g \to D_g$ such that:

1) $\nu(\partial \mathcal{D}_g) = K \subset D_g$, for every vertex v of K we have $\nu^{-1}(v) = \{v_1, v_2\}$, and a small neighborhood $U(v_i) \subset S$ is mapped to a pair of *opposite edges* of K at v;
2) the map ν is one-to-one everywhere except a union of *intervals*:
   $\Sigma = \{x \in D_g \,|\, \mathrm{card}(\nu^{-1}(x)) > 1\}$;
3) the set $\Sigma_3 = \{x \in D_g \,|\, \mathrm{card}(\nu^{-1}(x)) > 2\} \subset \Sigma$ is a finite subset of the zero-dimensional skeleton of the complex D_g and consists only of those points having exactly three preimages; moreover, $\Sigma_3 \cap \partial D_g = \emptyset$;
4) "local three-dimensionality": in the complex D_g neighborhoods of double points, triple points and cusps have the following structure:

   (a) a neighborhood of a cusp (points in which a map is not immersion) is homeomorphic the Whitney umbrella;
   (b) a neighborhood of a double point from $\Sigma \setminus \Sigma_3$ being not cusp is homeomorphic to a neighborhood of the point $(0,0,0)$ in the set $\{(x,y,z) \,|\, xy = 0\} \subset \mathbb{R}^3$;
   (c) a neighborhood of a triple point is homeomorphic to a neighborhood of the point $(0,0,0)$ in the set $\{(x,y,z) \,|\, xyz = 0\} \subset \mathbb{R}^3$;
   (d) a neighborhood of a vertex of K in D_g is homeomorphic to a neighborhood of the point $(0,0,0)$ in the set $\{(x,y,z) \,|\, xy = 0,\, z \geqslant 0\} \subset \mathbb{R}^3$.

The surface $\mathcal{D}_g$ will be called *the spanning surface* of genus g or the *cobordism of genus g* for K.

In other words, in the definition we require that a free knot (a framed 4-graph K represented by an image of a circle S) is spanned by the 2-complex D_g: an image of the 2-surface $\mathcal{D}_g$, the boundary of which is the circle S, and the singularities of the map $\nu \colon \mathcal{D}_g \to D_g$ are generic singularities (i.e., a neighborhood of each singularity is embedded in $\mathbb{R}^3$).

Analogously, one defines the slice genus for framed 4-graphs with many unicursal components, the cobordism of genus g for free knots (in this case a spanning surface has several boundary components), and the equivalence relation $\sim$ for free knots to be cobordant: two free knots K_1 and K_2 are *cobordant* if the free link $K_1 \sqcup K_2$ is cobordant to the unknot (with a cobordism of genus 0).

Remark 5.4. Further, saying a word "cobordism" we always mean a cobordism of genus 0 unless otherwise is indicated.

The closure $\overline{\Sigma}$ except for points from Σ, contains also *cusps*, i.e., such points $x \in D_g$ for which $\text{card}(\nu^{-1}(x)) = 1$, and, moreover, for any small neighborhood $U(x)$ of x the intersection $U(x) \cap \Sigma$ represents a punctured interval. Denote by Σ_2 the complement $\Sigma \setminus \Sigma_3$.

The intersection $\nu(S) \cap D_g$ is a framed 4-graph in D_g. This graph is obtained from $S = \partial D_g$ by *pasting double points* in S; the framing (the structure of opposite edges) for this graph is obtained from S: for a point x in $\nu(S) \cap \Sigma$ the preimage $\nu^{-1}(U(x) \cap \nu(S))$ consists of two branches of S; the images of those two branches under ν will generate the two pairs of opposite edges.

Definition 5.7. A free knot is called *null-cobordant* or *slice* if it has slice genus zero.

If a free knot K admits any cobordism of genus g and does not admit a cobordism of genus $g - 1$, we say that K has *slice genus g*. Notation: $sg(K) = g$.

The following lemma follows from the definition of a free knot:

Lemma 5.1. *If framed 4-graphs K, K' represent the same free knot, then K and K' are cobordant and, therefore, $sg(K) = sg(K')$.*

Indeed, in Fig. 5.4 we demonstrate that an equivalence under each Reidemeister move leads to them to be cobordant; since to be cobordant is an equivalence relation, we get the necessary. The first Reidemeister move corresponds to a cusp point, the second Reidemeister move corresponds to a passage through a tangency point, and the third Reidemeister move corresponds to a triple point.

Thus, it makes sense to speak about *the slice genus* of free knots, but not only for framed 4-graphs.

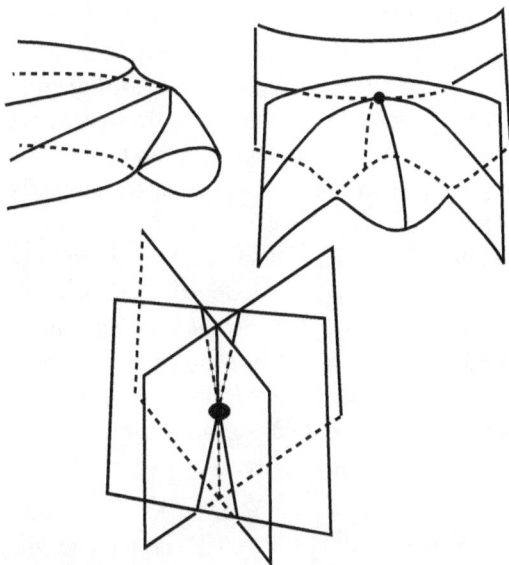

Fig. 5.4 Cobordisms corresponding Reidemeister moves

Remark 5.5. Let K be a flat knot and $|K|$ be the underlying free knot. Then it follows from the definition that the slice genus of K is greater than or equal to the slice genus of $|K|$. In particular, if K is slice, then so is $|K|$.

Example 5.2. The first example of a non-slice flat virtual knot was constructed by Carter [6], it is shown in Fig. 5.5.

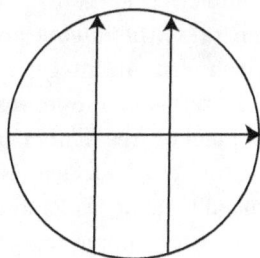

Fig. 5.5 Carter's non-slice flat virtual knot

This knot is embedded in an oriented surface of genus 2. Let us orient this surface. In Fig. 5.5, the arrows indicate the *clockwise direction of branches*. Namely, orient the core circle of the chord diagram counterclockwise and orient the immersed curve accordingly. If two oriented branches (a, b) of the curve have an intersection at a double point X and the tangent vectors $v_{X,a}$, $v_{X,b}$ form a positively-oriented basis, then the arrow is directed from a to b.

In this notation, for flat virtual knots, two chords participating in a second Reidemeister move should have opposite orientations. The flat knot K in Fig. 5.5 is non-trivial as a *flat virtual knot*, and, moreover, it is non-slice. Nevertheless, when we forget about the arrows, we can first cancel the two vertical arrows (by the second Reidemeister move) and then cancel the horizontal arrow (by a first Reidemeister move). So, the corresponding free knot $|K|$ is trivial, and hence *slice*.

So, if a free knot is *non-slice*, then so is *every* underlying virtual knot.

The problem of finding *non-slice* free knots is rather complicated.

In Sec. 5.2 we introduced the notion of combinatorial cobordism. Our combinatorial transformations consist of the third Reidemeister move and addition/removal of an "even symmetric configuration". The latter transformations include the first and second Reidemeister moves as particular cases, and each of them represents a (topological) cobordism of genus zero. Thus, if two knots are combinatorially cobordant, then they are also topologically cobordant.

In the work by Carter [6] and Turaev [67], topological sliceness obstructions for immersed curves (which come from flat knots) were studied. For each double point x of an immersed curve Γ, one considers the homology class of the halves $\Gamma_{x,1}$, and takes the *homological pairing* of these halves in the surface. These pairings form an integer matrix, an obstruction for a cobordism is formulated in terms of properties of the obtained matrix. This approach cannot be applied to free knots because a framed 4-graph is not assumed to be embedded in *any* 2-surface. Moreover, embeddings into different 2-surfaces may crucially change the intersection form for "halves" even with $\mathbb{Z}_2$–coefficients. As an example one can consider the standard diagram of the trefoil and a diagram of the trefoil after virtualising one of these crossings (this diagram lies in the torus).

5.5 Parity of curves in 2-surfaces

Let us now pay more attention to the structure of cobordisms of free knots. Assume there is a cobordism $\nu \colon \mathcal{D} \to D$ (of genus zero) spanning the free knot (framed 4-graph) $K = \nu(\partial \mathcal{D})$.

Set $\Psi = \nu^{-1}(\overline{\Sigma})$. Then Ψ has a natural stratification containing strata of dimensions zero and one. The strata of dimension 0 are *double points on the boundary*, *cusps*, and *triple points*; all other points form strata of dimension one. By a *double line* we mean a minimal (with respect to inclusion) collection of 1-dimensional strata possessing the following properties:

1) two 1-strata attaching the same cusp from opposite sides belong to the same double line;
2) two 1-strata attaching the same triple point from *opposite* sides belong to the same double line (see Fig. 5.6).

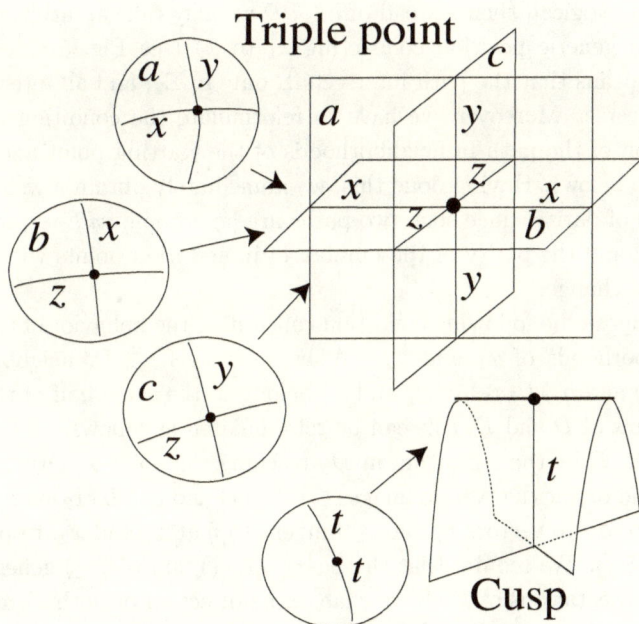

Fig. 5.6   Double lines a, x, y, z, cusps and triple points

Let $x \in K \cap \Sigma$ be a double point on the boundary of ∂D. Assume $\nu^{-1}(x) = \{x_1, x_2\}$.

Recall the definition of the Gaussian parity for framed 4-graphs. Let us consider a double point x: we take an arc (a half of the core circle) ι of K connecting x to x, being the image of an arc $\tilde{\iota}$ on S, and count the parity of the double point number on ι. Note that here we take a half of the core circle, but not just a curve on K connecting the point x to itself. The fact of the matter is that if we take a curve on the circle $\partial \mathcal{D}$ connecting two preimages x_1 and x_2 of x, but locally (in neighborhoods of these preimages) pointed towards *different* "halves", then the number of double points on such curve is greater (or smaller) by one, since one of the preimages of x is added.

Now, let us consider the preimage $\tilde{\iota} \subset S$ connecting x_1 to x_2. Then the definition of parity $p(x)$ can be reformulated as the parity of $\mathrm{card}(\tilde{\iota} \cap \Sigma)$. Note that this line $\tilde{\iota}$ belongs to the disc $\mathcal{D}$.

If we want to generalise the notion of parity on double lines and make it more topological, then instead of $\tilde{\iota} \subset \partial D$ we may take an arbitrary path $\tilde{\eta} \subset \mathcal{D}$ in generic position connecting x_1 to x_2 (see Fig. 5.7). Generic position means that the path intersects $\overline{\Sigma}$ only in Σ_2, and all intersections are transverse. Moreover, we have to reformulate the condition of right-orientation of the path in neighborhoods of the starting point and ending point (see below). Having done this, we immediately obtain a well-defined definition of parity, since such two paths are homotopic *with respect to the boundary*, and the parity of the number of intersection points with the set Σ will not change.

We impose the following condition concerning the behavior of the curve in neighborhoods of x_1 and x_2. When we take $\tilde{\eta} \subset \mathcal{D}$, neighborhoods $\tilde{\eta} \cap U(x_1)$ and $\tilde{\eta} \cap U(x_2)$ of x_1 and x_2 belong to the same half of the circle S. In terms of $\mathcal{D}$ and D, this can be reformulated as follows.

Let $\zeta \subset D$ be the 1-stratum in D attaching the point x. Orient ζ arbitrarily, and orient the two preimages $\zeta_1 \cap U(x_1)$ and $\zeta_2 \cap U(x_2)$ accordingly. Consider the two vectors v_1 and v_2 tangent to $\tilde{\eta}$ at x_1 and x_2, respectively (see Fig. 5.7). We require that the bases $(\dot{\zeta}_1, v_1)$ and $(\dot{\zeta}_2, v_2)$ generate two *different* orientations of $\mathcal{D}$. If we change the direction of both x_1 and x_2 it will not change the parity of $\tilde{\eta} \cap \Sigma_2$.

So, this gives another way to define a parity for x. Consider it as the

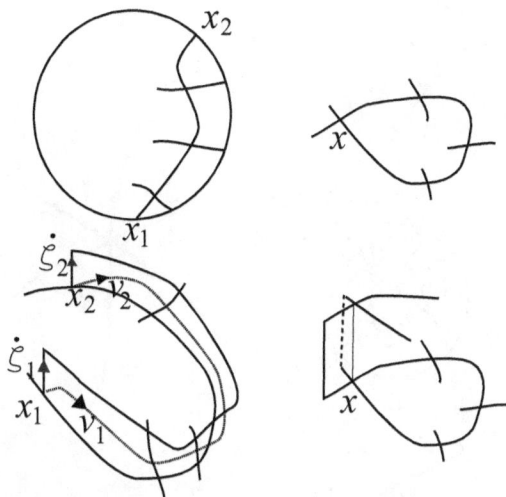

Fig. 5.7 The geometric way for defining parity

definition of parity for *any* double point from Σ_2.

Actually, let x be a point on a double line from Σ_2. Having counted the number of intersection points of a curve connecting x_1 to x_2 by the way mentioned above, we get the number $p(x)$. We call it *the Gaussian parity* of the point x.

The following statement easily follows from the definition

Statement 5.4. *The Gaussian parity is constant along double lines.*

Proof. This is evident for two points belonging to the same 1-stratum and for points on two 1-strata attaching the same cusp. When passing through a triple point, the parity does not change (see Fig. 5.8). We see that the curve connecting the two preimages of A is "parallel" to the curve connecting the two preimages of B everywhere except for the two small domains; inside these two domains, we have two intersections with double lines p and q which cancel each other. Thus, the parity of the points A and B coincide. $\qquad\square$

This leads to the definition of parity for a *double line*:

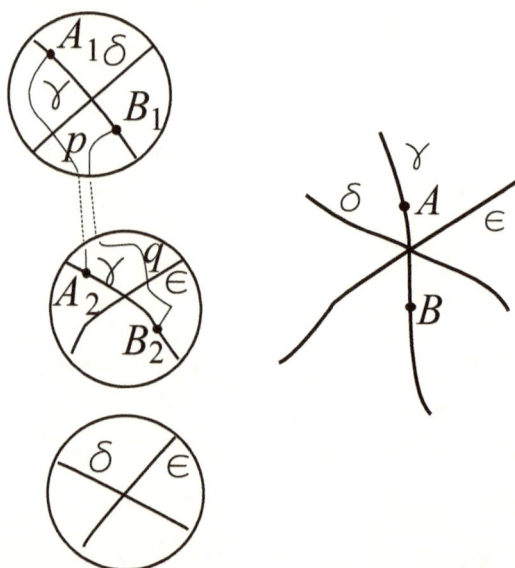

Fig. 5.8   The behavior of parity and justified parity when passing through a triple point

Definition 5.8. Let γ be a double line on a cobordism. Take an arbitrary x on $\gamma \cap \Sigma_2$, and consider the two preimages x_1 and x_2 of it on $\mathcal{D}$. Connect x_1 to x_2 by a generic path $\tilde{\eta}$ such that its behavior in neighborhoods $U(x_1)$ and $U(x_2)$ is coordinated (as in the definition of the Gaussian parity). Now, the *Gaussian parity* of the double line containing x is the parity of the number of intersection points between $\tilde{\eta}$ and Ψ.

This allows us to define the set Ψ_{even} (respectively, Ψ_{odd}) to consist of the closure of all those points of Ψ belonging to *even* (respectively, *odd*) double lines.

From the definition of the Gaussian parity for double lines it is easy to get the following statement.

Statement 5.5. *Amongst three double lines intersecting in the same point the number of odd lines is even (equals zero or two).*

Now, to define the *justified Gaussian parity*, we should take an *odd double line* γ, an arbitrary generic point x on it, and consider the two preimages x_1 and x_2 of x. Then we connect x_1 to x_2 by a generic path $\delta \in \mathcal{D}$ (of course, its behavior in neighborhoods $U(x_1)$ and $U(x_2)$ is coordinated) and count the number of intersections between δ and Ψ_{even}. If this number is even, we say that the 1-stratum containing x is of the *first type*; otherwise, we say that this 1-stratum is of the *second type*.

Note that here we do not require any coordination (coorientation) of the first and last segments of the path: under small perturbation of the path in a neighborhood of its start (end) the number of intersection points only with *odd double line* can change, but we are interested in intersections with even double lines.

From the definition it immediately follows statement.

Statement 5.6. *The Gaussian justified parity is constant on 1-strata belonging to Ψ. It does not change when passing through a cusp point (from one stratum to another stratum of the same double line), and it changes from b to b' or from b' to b when passing through a triple point formed by two odd double lines and one even double line.*

Proof. The statement that the type does not change along 1-strata is evident. Consider now Fig. 5.8. Assume the double lines γ and δ are odd and the double line ϵ is even. Then the two lines connecting the preimages of A and the preimages of B are "parallel" except for two domains where one of them passes through a double line at p, and the other one passes through a double line at q. Since we disregard intersection with *odd* double lines, we see that q counts and p does not. Thereby, we have proved that the type changes by 1 when we pass through a triple point. $\quad\square$

Having defined the *Gaussian parity* for double lines and the *Gaussian justified parity* for strata, we shall be able to construct the invariant L for any *section* of a cobordism D. The parity properties proved above for the parity of double lines and the justified parity of strata guarantee that this invariant is well defined and behaves nice under the Reidemeister moves.

5.6   Sliceness of free knots

It turns out that the invariant L of free knots is an *obstruction to sliceness*.

Before proving the main theorem, we shall make several observations concerning sliceness. From Lemma 5.1 it follows that the slice genus is well defined on the set of free knots.

The following statement is trivial (see Fig. 5.4).

Statement 5.7. *If a framed 4-graph K' is equivalent to a framed 4-graph K (by Reidemeister moves), then the slice genus of K' is equal to that of K. In particular, if K is slice, then so is K'.*

Thus, it makes sense to speak about *cobordisms* of free knots, but not only for framed 4-graphs.

As a corollary, we get the following statement.

Statement 5.8. *If a framed 4-graph K is embeddable in S^2 or T^2, then K is slice.*

Indeed, every framed 4-graph on S^2 is equivalent (even as a flat virtual knot) to the (free) unknot; every framed 4-graph on the torus T^2 is homotopic either to the trivial loop on the torus (with no crossings) or to a knot lying in a cylinder (a complement to a simple non-contractible curve in the torus). In the latter case, this free knot is equivalent to a knot lying in a sphere (as cylinder is a submanifold of a sphere) and, therefore, it is trivial.

Statement 5.9. *Let K be a free knot, and let $f(K)$ be a free knot obtained from K by deleting odd crossings. If K is slice, then so is $f(K)$.*

Proof. Indeed, any cobordism (of genus zero) for K generates a cobordism of genus zero for $f(K)$ obtained by separating all *odd double lines*: two points from Σ_2 will be pasted together in the new cobordism only if and only if they lie on an even double line in the first cobordism. Note that the rule of pasting is agreed with triple points, since if we have two even double lines in a triple point, then the third double line is even too. $\square$

Denote the obtained cobordism for $f(K)$ by $f(D)$, where D is the cobordism for K.

The main result of the section is the following theorem.

Theorem 5.3. *If a free knot K has $L(K) \neq 0$, then K is not slice.*

In particular, this yields the following

Corollary 5.4. *Let γ be a curve immersed in an oriented closed 2-surface S_g. Then if for a free knot Γ corresponding to γ one has $L(\Gamma) \neq 0$, then the flat virtual knot corresponding to γ is not slice.*

Indeed, a disc immersed in a 3-manifold is a spanning disc. The converse statement is, generally, not true.

The problem of finding obstructions for a surface S_g with a curve γ to span a disc immersed in a 3-manifold M with boundary S_g was studied by Carter [6], Turaev [67] etc. Some topological obstructions based on homology of S_g were constructed.

In the present section we are considering a more complicated problem: instead of curves in 2-surfaces we consider framed 4-graphs, and instead of spanning 2-discs in 3-manifolds we consider "abstract" spanning 2-discs. In this case, we cannot define a "homology group", since a framed 4-graph can be embedded in different surfaces, which have different homology groups.

From this point of view the notion of parity plays in some sense the role of a "substitute for homology of S_g".

The proof of the main theorem will consist of several steps.

First, let us adopt the following notation: by ν we shall denote a map $\mathcal{D} \to D$ corresponding to the cobordism, and by μ we shall denote as a Morse function $\mu \colon D \to \mathbb{R}$ on the disc with self-intersections (see the definition below) as the composition $\mu \circ \nu \colon \mathcal{D} \to \mathbb{R}$ representing a Morse function on the disc $\mathcal{D}$.

Assume a free knot K (represented by a framed 4-graph) admits a cobordism $\nu \colon \mathcal{D} \to D$ (of genus zero).

Definition 5.9. By a *Morse function* on D we mean a Morse function $\mu \colon \mathcal{D} \to [0, \infty)$ such that if $\nu(x) = \nu(y)$, then $\mu(x) = \mu(y)$, all triple points and cusp points on $\mathcal{D}$ lie on non-critical levels of μ, and $\mu^{-1}(0) = K$, $\mu^{-1}(1) = \emptyset$. By abuse of notation, we shall denote the function on $\mathcal{D}$ and the function on D by the same letter f.

By a *non-singular* value of the function μ we mean a non-critical value X of the function μ such that $\mu^{-1}(X) \subset \mathcal{D}$ contains no cusps and no triple points; a singular (respectively, non-singular) level is the preimage of a singular (respectively, non-singular) value. A Morse function on D will be called *simple* if every singular level contains either exactly one critical point, or exactly one triple point, or exactly one cusp point.

From now on, we require that the Morse function on D is simple and the level 0 is non-singular. It is clear that such Morse functions are *everywhere dense in the class of all functions*. Every Morse function has singular levels of two types: those corresponding to Morse bifurcations (saddles, minima, and maxima) and those corresponding to Reidemeister moves. Denote singular levels of the Morse function μ by $c_1 < \ldots < c_k$ and choose non-singular levels a_i: $0 = a_0 < c_1 < a_1 < c_2 < \ldots < a_k < c_k < a_{k+1} = 1$.

Let us construct the *Reeb graph* Γ_μ (*molecule*) of the Morse function μ as follows. All vertices of this graph have degree either 1 or 3. The univalent vertices of the Reeb graph (except one) will correspond to minima and maxima of the function μ; the vertices of valence three will correspond to saddle points (note that each saddle consists of a transformation of one circle to two circles or two circles to one, since the surface–disc is oriented); edges will connect critical points; every edge will correspond to a cylinder $S^1 \times I \subset D$ which is continuously mapped by μ to a closed interval (this edge) between some two vertices (critical points); this cylinder has no critical Morse points inside (but may contain inside critical points corresponding to Reidemeister moves). One edge will emanate from the point 0 (a non-critical point) corresponding to the circle $S = \partial \mathcal{D}$ (see Fig. 5.9).

Since this graph is the Reeb graph of a Morse function on the disc $\mathcal{D}$, the graph Γ_μ is a *tree*.

Our next goal is to endow each edge of the Reeb graph with a non-negative integer *label*. The label of the edge emanating from 0 will coincide with $L(K)$.

For every non-singular level c of μ the preimage $K_c = \mu^{-1}(c) \subset D$ is a framed 4-graph representing a free link; when passing through a Reidemeister singular point, it is operated on by the corresponding Reidemeister move; when passing through a Morse critical point it gets operated on by a Morse-type bifurcation. Every crossing of K_c belongs to some double line. Define the *parity* of a crossing to be the Gaussian parity of the double line, it belongs to. Analogously, define the *justified parity* of a crossing to be that of the 1-stratum, it belongs to.

One can easily see that if a section of the Morse function is a free

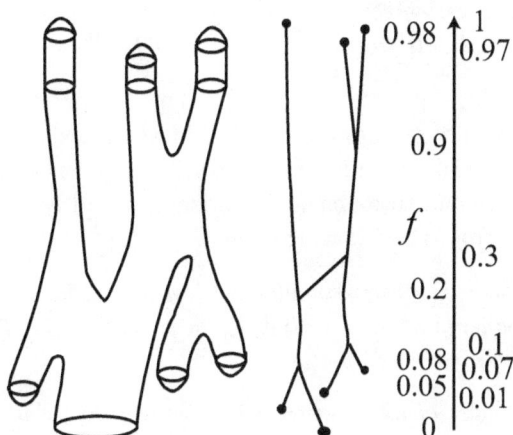

Fig. 5.9   The Reeb graph and circle bifurcations

knot (a framed 4-graph with one unicursal component), then the parity and justified parity coincide with the Gaussian parity and justified parity defined directly via Gauss diagrams.

Choose a non-singular level c, and consider the free link K_c and orient its component arbitrarily (as we shall see further, the orientation will be immaterial); for every unicursal component $K_{c,j}$ of the free link K_c we may define the conjugacy class $\delta(K_{c,j})$ of $\gamma(K_{c,j})$ in G just as it is done for free knots with the Gaussian parity and justified parity. Let $\delta(K_c)$ be the unordered collection of all $\delta(K_{c,j})$ for all j (with repetitions).

From Statements 5.4 and 5.6 we get the following statement.

Lemma 5.2. *The parity and justified parity defined on the set of all non-singular levels K_c satisfy the parity and justified parity axioms under those Reidemeister moves which happen under passing from one non-singular level to another one within the cobordism D.*

Thus, we get the following lemma.

Lemma 5.3. *When changing a parameter c along an interval $[a, b] \subset [0, 1]$ on which the level K_c is operated on by Reidemeister moves but not Morse bifurcations, and the levels corresponding to a and b contain no Reidemeister move, we have $\delta(K_a) = \delta(K_b)$ if the orientations of components of the* links K_a and K_b are agreed with each other: these sets represent conjugacy classes of elements from G. Moreover, $\delta(K_0)$ is the conjugacy class of $\gamma(K)$.

The proof literally repeats the proof of Theorem 5.2.

Now, we would like to treat δ as a collection of *non-negative integers* with multiplicities and to forget about orientations of components of K_c. To this end, we prove the following lemma.

Lemma 5.4. *Let c be a non-singular level of μ, and let $K_{c,1}, \ldots, K_{c,n}$ be unicursal components of the free link $K_c = \mu^{-1}(c) \subset D$. Then for every $i = 1, \ldots, n$ the following properties hold:*

1) *the total number of intersection points between $K_{c,i}$ and $K_{c,j}$, $j \neq i$, is even;*

2) *the number of odd intersection points between $K_{c,i}$ and $K_{c,j}$, $j \neq i$, (i.e., intersection points lying on odd double lines) is even.*

Proof. The proof follows from the fact that the preimage of the framed 4-graph $K_{c,i}$ in $\mathcal{D}$ is a circle, and the intersection of a closed curve with the set Ψ (or Ψ_{even}) in $\mathcal{D}$ consists of an even number of points. $\square$

Lemma 5.4 immediately means that every $\gamma(K_{c,i})$ for a non-singular value c, is represented by an element $(0, 2k)$ on the Cayley graph of G (for some integer k).

Let $L_c = \{l_{c,1}, \ldots, l_{c,m}\}$ be the unordered collection of integers (with repetitions) obtained from $\delta(K, c)$ by replacing conjugacy classes of elements from G with absolute values of their second coordinates. Since the reversing of the orientation changes the coordinate of an element of the group G from $(0, 2k)$ to $(0, -2k)$, then under reversing the orientation the unordered collection of modules of numbers (which is talked about in Lemma 5.4) corresponding to components of links, does not change.

Each $l_{c,i} \in \mathbb{N} \cup \{0\}$ corresponds to a component of the free link K_c and does not change under Reidemeister moves when changing c without passing through Morse critical points. Associate it with the corresponding edge of the graph Γ_μ.

Now, let us analyze the behavior of these labels $l_{c,i}$ at vertices of the graph Γ_μ.

Lemma 5.5. *Assume $K_{c-\varepsilon}$ and $K_{c+\varepsilon}$ differ by one Morse bifurcation at the level c. Then:*

1) *If this bifurcation corresponds to a birth of a circle, then $L_{c+\varepsilon}$ is obtained from $L_{c-\varepsilon}$ by an addition of 0.*
2) *If it corresponds to a removal of a circle, then $L_{c+\varepsilon}$ is obtained from $L_{c-\varepsilon}$ by a removal of 0.*
3) *In the case of fusion of two circles into one, the set $L_{c+\varepsilon}$ is obtained from $L_{c-\varepsilon}$ by applying the following operation: all elements except two (being equal to m and n) remain the same, and the elements m and n turn into some $k = \pm n \pm n$ to form an element.*
4) *The fission operation is the inverse to the fusion: instead of one element k one gets a pair of elements m, n such that $\pm m \pm n = k$.*

Proof. The first two assertions are obvious: the trivial circle has no double points, thus the corresponding element of G is the unit of G, its coordinate is $(0,0)$ and the corresponding label is equal to 0.

The last two assertions follow from the following observation. If a circle C with a marked point X splits into two circles by a Morse bifurcation connecting X to some point Y, then the corresponding word $w \in G$ splits into the product $w = w_{XY} w_{YX}$.

The rest of the proof follows from the multiplication rule in G: for elements u, $v \in G$ having coordinates (u_1, u_2) and (v_1, v_2), respectively, the product $u \cdot v$ has coordinates $(\pm u_1 \pm v_1, \pm u_2 \pm v_2)$. $\qquad\square$

The proved lemma leads to the following way of proving Theorem 5.3. The graph Γ_μ has all vertices of degree one except possibly one (corresponding to the initial knot K), having label 0. At each vertex of degree 3 the three labels with signs $\pm$ sum up to give zero. Thus, taking into account that the Reeb graph is a tree, we get $L(K) = 0$. The contradiction completes the proof of Theorem 5.3.

Example 5.3. Consider the free knot K_1 shown in Fig. 5.3. By Theorem 5.3, it is not-slice. Thus, all flat virtual knots K with the underlying free knot K_1, are not slice, either.

Bibliography

[1] Afanas'ev, D. M. (2010). Refining virtual knot invariants by means of parity, *Sb. Math.* **201**, 6, pp. 785–800, *Mat. Sb.* **201**, 6, pp. 3–18 (in Russian).

[2] Bourgoin, M. O. (2008). Twisted link theory, *Algebr. Geom. Topol.* **8**, 3, pp. 1249–1279.

[3] Brailov, A. V. and Fomenko, A. T. (1989). The topology of integral submanifolds of completely integrable Hamiltonian systems, *Math. USSR-Sb.* **62**, 2, pp. 373–383.

[4] Cairns, G. and Elton, D. (1993). The planarity problem for signed Gauss words, *J. Knot Theory Ramifications* **2**, pp. 359–367.

[5] Cairns, G. and Elton, D. (1996). The planarity problem. II, *J. Knot Theory Ramifications* **5**, pp. 137–144.

[6] Carter, J. S. (1991). Closed curves that never extend to proper maps of disks, *Proc. Amer. Math. Soc.* **113**, 3, pp. 879–888.

[7] Carter, J. S., Kamada, S. and Saito, M. (2002). Stable equivalence of knots on surfaces, *J. Knot Theory Ramifications* **11**, pp. 311–322.

[8] Cerf, J. (1968). *Sur les difféomorphismes de la sphère de dimension trois* ($\Gamma_4 = 0$), Lecture Notes in Mathematics, Vol. 53, Berlin: Springer-Verlag, xii+133 pp.

[9] Chrisman, M. (2014). Knots in virtually fibered 3-manifolds, preprint, arXiv:math.GT/1405.6072.

[10] Chrisman, M. and Manturov, V. O. (2010). Combinatorial formulae for finite-type invariants via parities, preprint, arXiv:math.GT/1002.0539.

[11] Dye, H. A. and Kauffman, L. H. (2009). Virtual crossing number and the arrow polynomial, *J. Knot Theory Ramifications* **18**, 10, pp. 1335–1357.

[12] Dye, H. A., Kauffman, L. H. and Manturov, V. O. (2010). On two categorifications of the arrow polynomial for virtual knots, in *The Mathematics of Knots*, Contributions in the Mathematical and Computational Sciences, Vol. 1, Berlin: Springer, pp. 95–124.

[13] Fomenko, A. T. (1986). Morse theory of integrable Hamiltonian systems,

Soviet Math. Dokl. **33**, 2, pp. 502–506.

[14] Fomenko, A. T. (1987). The topology of surfaces of constant energy in integrable Hamiltonian systems, and obstructions to integrability, *Math. USSR Izv.* **29**, 3, pp. 629–658.

[15] Fomenko, A. T. (1988). Topological invariants of Hamiltonian systems that are integrable in the sense of Liouville, *Funct. Anal. Appl.* **22**, 4, pp. 286–296.

[16] Fomenko, A. T. (1989). The symplectic topology of completely integrable Hamiltonian systems, *Russian Math. Surveys* **44**, 1, pp. 181–219.

[17] Fomenko, A. T. (1991). The theory of invariants of multidimensional integrable hamiltonian systems (with arbitrary many degrees of freedom). Molecular table of all integrable systems with two degrees of freedom, *Adv. Sov. Math.* **6**, pp. 1–35.

[18] Gibson, A. (2009). Homotopy invariants of Gauss words, preprint, arXiv:math.GT/0902.0062.

[19] A. Gibson, PhD thesis.

[20] Goldman, W. (1986). Invariant functions on Lie groups and Hamiltonian flows of surface group representations, *Invent. Math.* **85**, pp. 263–302.

[21] Gordon, C. McA. and Luecke, J. (1989). Knots are determined by their complements, *J. Amer. Math. Soc.* **2**, 2, pp. 371–415.

[22] Goussarov, M., Polyak, M. and Viro, O. (2000). Finite type invariants of classical and virtual knots, *Topology* **39**, pp. 1045–1068.

[23] Habegger, N., Lin, X.-S.X.-S. (1990). The Classification of Links up to Link-Homotopy, *J. Amer. Math. Soc.* **3**, 2, pp. 389–419.

[24] Hass, J. and Scott, P. (1994). Shortening curves on surfaces, *Topology* **33**, 1, pp. 25–43.

[25] Ilyutko, D. P. and Manturov, V. O. (2009). Introduction to graph-link theory, *J. Knot Theory Ramifications* **18**, 6, pp. 791–823.

[26] Ilyutko, D. P. and Manturov, V. O. (2009). Graph-links, *Dokl. Math.* **80**, 2, pp. 739–742, *Dokl. Akad. Nauk* **428**, 5, pp. 591–594 (in Russian).

[27] Ilyutko, D. P. and Manturov, V. O. (2009). Cobordisms of free knots, *Dokl. Math.* **80**, 3, pp. 1–3, *Dokl. Akad. Nauk* **429**, 4, pp. 439–441 (in Russian).

[28] Ilyutko, D. P., Manturov, V. O. and Nikonov, I. M. (2013). Parity in knot theory and graph-links, *J. Math. Sc.* **193**, 6, pp. 809–965.

[29] Ilyutko, D. P., Manturov, V. O. and Nikonov, I. M. (2014). Virtual knot invariants arising from parities, *Banach Center Publ.* **100**, pp. 99–130.

[30] Jones, V. F. R. (1985). A polynomial invariant for links via Neumann algebras, *Bull. Amer. Math. Soc.* **129**, pp. 103–112.

[31] Kamada, N. and Kamada, S. (2000). Abstract link diagrams and virtual knots, *J. Knot Theory Ramifications* **9**, 1, pp. 93–109.

[32] Kauffman, L. H. (1997). Virtual knots, *talks at MSRI Meeting*, January 1997 and AMS meeting at University of Maryland, College Park, March 1997.

[33] Kauffman, L. H. (1999). Virtual knot theory, *European J. Combin.* **20**, 7,

pp. 663–690.

[34] Kauffman, L. H. (2004). A self-linking invariant of virtual knots, *Fund. Math.* **184**, pp. 135–158.

[35] Kauffman, L. H. and Manturov, V. O. (2006). Virtual knots and links, *Proc. Steklov Inst. Math.* **252**, pp. 104–121, *Tr. Mat. Inst. Steklova* **252**, pp. 114–133 (in Russian).

[36] Krasnov, V. A. and Manturov, V. O. (2013). Graph-valued invariants of virtual and classical links and minimality problem, *J. Knot Theory Ramifications* **22**, 12, 1341006.

[37] Krylov, D. Yu. and Manturov, V. O. (2011). Parity and relative parity in knot theory, preprint, arXiv:math.GT/1101.0128.

[38] Kuperberg, G. (2003). What is a virtual link? *Algebr. Geom. Topol.* **3**, pp. 587–591.

[39] Lowrance, A. (2007). Heegaard–Floer homology and Turaev genus, preprint, arXiv:math.GT/0709.0720.

[40] Manturov, O. V. and Manturov, V. O. (2009). Free knots and groups, *J. Knot Theory Ramifications* **18**, 2, pp. 181–186.

[41] Manturov, V. O. (1998). Atoms, height atoms, chord diagrams, and knots. Enumeration of atoms of low complexity using Mathematica 3.0, in *Topological Methods in Hamiltonian Systems Theory*, Moscow: Factorial, pp. 203–212 (in Russian).

[42] Manturov, V. O. (2000). The bracket semigroup of knots, *Math. Notes* **67**, 4, pp. 468–478, *Mat. Zametki* **67**, 4, pp. 549–562 (in Russian).

[43] Manturov, V. O. (2000). Bifurcations, atoms and knots, *Moscow Univ. Math. Bull.* **55**, 1, pp. 1–7, *Vestnik Moskov. Univ. Ser. I Mat. Mekh* **1**, pp. 3–8 (in Russian).

[44] Manturov, V. O. (2005). *Teoriya uzlov (Knot theory)*, Moscow–Izhevsk: RCD, 512 pp. (in Russian).

[45] Manturov, V. O. (2005). On long virtual knots, *Dokl. Math.* **71**, 2, pp. 253–255, *Dokl. Akad. Nauk* **401**, 5, pp. 595–598 (in Russian).

[46] Manturov, V. O. (2005). A proof of Vassiliev's conjecture on the planarity of singular links, *Izv. Math.* **69**, 5, pp. 1025–1033, *Izvestiya RAN, Ser. Mat.* **69**, 5, pp. 169–178 (in Russian).

[47] Manturov, V. O. (2009). On free knots, preprint, arXiv:math.GT/0901.2214.

[48] Manturov, V. O. (2009). On free knots and links, preprint, arXiv:math.GT/0902.0127.

[49] Manturov, V. O. (2009). Free knots are not invertible, preprint, arXiv:math.GT/0909.2230v2.

[50] Manturov, V. O. (2010). Parity in knot theory, *Sb. Math.* **201**, 5, pp. 693–733, *Mat. Sb.* **201**, 5, pp. 65–110 (in Russian).

[51] Manturov, V. O. (2011). Parity, free knots, groups, and invariants of finite type, *Trans. Moscow Math. Soc.*, pp. 157–169, *Tr. Mosk. Mat. Obs.* **72**, 2,

pp. 207–222 (in Russian).

[52] Manturov, V. O. (2012). Parity and cobordisms of free knots, *Sb. Math.* **203**, 5, pp. 196–223, *Math. sb.* **203**, 2, pp. 45–76 (in Russian).

[53] Manturov, V. O. (2012). Free knots and parity, in *Introductory Lectures on Knot Theory, Selected Lectures Presented at the Advanced School and Conference on Knot Theory and its Applications to Physics and Biology, Series of Knots and Everything*, Vol. 46, World Scientific, pp. 321–345.

[54] Manturov, V. O. (2012). A fuctorial map from knots in thickened surfaces to classical knots and generalisations of parity, preprint, arXiv:math.GT/1011.4640v4.

[55] Manturov, V. O. (2013). Parity and projection from virtual knots to classical knots, *J. Knot Theory Ramifications* **22**, 1350044.

[56] Manturov, V. O. and Ilyutko, D. P. (2013). *Virtual Knots: The State of the Art*, Singapore: World Scientific, 547 pp.

[57] Milnor, J. (1954). Link Groups, *Ann. of Math.* **59**, pp. 177–195.

[58] Miyazawa, Y. (2006). Magnetic graphs and an invariant for virtual links, *J. Knot Theory Ramifications* **15**, 10, pp. 1319–1334.

[59] Polyak, M. and Viro, O. (1994). Gauss diagram formulae for Vassiliev invariants, *Int. Math. Res. Not.* **11**, pp. 445–453.

[60] Read, R. C. and Rosenstiehl, P. (1976). *On the Gauss Crossing Problem*, Colloq. Math. Soc. Janos Bolyai, Amsterdam: North-Holland, pp. 843–876.

[61] Reidemeister, K. (1932). *Knotentheorie*, Berlin: Springer, 74 pp.

[62] Satoh, S. (2000). Virtual knot presentation of ribbon torus-knots, *J. Knot Theory Ramifications* **9**, 4, pp. 531–542.

[63] Stoimenov, A., Tchernov, V. and Vdovina, A. (2002). The canonical genus of a classical and virtual knot, *Geom. Dedicata* **95**, pp. 215–225.

[64] Turaev, V. G. (1987). A simple proof of the Murasugi and Kauffman theorems on alternating links, *Enseign. Math.* (2) **33**, 3–4, pp. 203–225.

[65] Turaev, V. G. (1991). Skein quantization of Poisson algebras of loops on surfaces, *Ann. Sci. École Norm. Sup.* **4**, 24, 635–704.

[66] Turaev, V. G. (2003). Virtual strings and their cobordisms, prerpint, arXiv:math.GT/0311185.

[67] Turaev, V. G. (2005). Cobordism of words, preprint, arXiv:math.CO/0511513.

[68] Turaev, V. G. (2005). Knots and words, preprint, arXiv:math.GT/0506390v1.

Index